CONFRONTING GLOBAL WARMING

# Population, Resources, and Conflict

CONFRONTING GLOBAL WARMING

# Population, Resources, and Conflict

Jacqueline Langwith

Michael E. Mann
*Consulting Editor*

**GREENHAVEN PRESS**
*A part of Gale, Cengage Learning*

Detroit • New York • San Francisco • New Haven, Conn • Waterville, Maine • London

Christine Nasso, *Publisher*
Elizabeth Des Chenes, *Managing Editor*

© 2011 Greenhaven Press, a part of Gale, Cengage Learning

For more information, contact:

Greenhaven Press
27500 Drake Rd.
Farmington Hills, MI 48331-3535
Or you can visit our Internet site at
gale.cengage.com.

For product information and technology assistance, contact us at **Gale Customer Support, 1-800-877-4253**.

For permission to use material from this text or product, submit all requests online at **www.cengage.com/permissions**.

Further permissions questions can be e-mailed to permissionrequest@cengage.com

Every effort is made to ensure that Greenhaven Press accurately reflects the original intent of the authors. Every effort has been made to trace the owners of copyrighted material.

Cover image © Dinodia Photos/Alamy.

**LIBRARY OF CONGRESS
CATALOGING-IN-PUBLICATION DATA**

Langwith, Jacqueline.
  Population, resources, and conflict / by Jacqueline Langwith.
    p. cm. -- (Confronting global warming)
  Includes bibliographical references and index.
  ISBN 978-0-7377-5175-8 (hbk.)
  1. Population--Environmental aspects. 2. Population. 3. Global warming. I. Title.
  HB849.415.L36 2011
  363.738'74--dc22
                                        2010050123

Printed in the United States of America
1 2 3 4 5 6 7 15 14 13 12 11

# Contents

# Preface

> "The warnings about global warming
> have been extremely clear for a long
> time. We are facing a global climate
> crisis. It is deepening. We are entering
> a period of consequences."
>
> *Al Gore*

Still hotly debated by some, human-induced global warming is now accepted in the scientific community. Earth's average yearly temperature is getting steadily warmer; sea levels are rising due to melting ice caps; and the resulting impact on ocean life, wildlife, and human life is already evident. The human-induced buildup of greenhouse gases in the atmosphere poses serious and diverse threats to life on earth. As scientists work to develop accurate models to predict the future impact of global warming, researchers, policy makers, and industry leaders are coming to terms with what can be done today to halt and reverse the human contributions to global climate change.

Each volume in the Confronting Global Warming series examines the current and impending challenges the planet faces because of global warming. Several titles focus on a particular aspect of life—such as weather, farming, health, or nature and wildlife—that has been altered by climate change. Consulting the works of leading experts in the field, Confronting Global Warming authors present the current status of those aspects as they have been affected by global warming, highlight key future challenges, examine potential solutions for dealing with the results of climate change, and address the pros and cons of imminent changes and challenges. Other volumes in the series—such as those dedicated to the role of government, the role of industry, and the role of the individual—address the impact various facets of society can have on climate change. The result is a series that provides students and general-interest readers with a solid understanding of the worldwide ramifications of climate change and what can be done to help humanity adapt to changing conditions and mitigate damage.

Each volume includes:

- A descriptive **table of contents** listing subtopics, charts, graphs, maps, and sidebars included in each chapter
- Full-color **charts, graphs, and maps** to illustrate key points, concepts, and theories
- Full-color **photos** that enhance textual material
- **Sidebars** that provide explanations of technical concepts or statistical information, present case studies to illustrate the international impact of global warming, or offer excerpts from primary and secondary documents
- **Pulled quotes** containing key points and statistical figures
- A **glossary** providing users with definitions of important terms
- An annotated **bibliography** of additional books, periodicals, and Web sites for further research
- A detailed **subject index** to allow users to quickly find the information they need

The Confronting Global Warming series provides students and general-interest readers with the information they need to understand the complex issue of climate change. Titles in the series offer users a well-rounded view of global warming, presented in an engaging format. Confronting Global Warming not only provides context for how society has dealt with climate change thus far but also encapsulates debates about how it will confront issues related to climate in the future.

# Foreword

Earth's climate is a complex system of interacting natural components. These components include the atmosphere, the ocean, and the continental ice sheets. Living things on earth—or, the biosphere—also constitute an important component of the climate system.

## Natural Factors Cause Some of Earth's Warming and Cooling

Numerous factors influence Earth's climate system, some of them natural. For example, the slow drift of continents that takes place over millions of years, a process known as plate tectonics, influences the composition of the atmosphere through its impact on volcanic activity and surface erosion. Another significant factor involves naturally occurring gases in the atmosphere, known as greenhouse gases, which have a warming influence on Earth's surface. Scientists have known about this warming effect for nearly two centuries: These gases absorb outgoing heat energy and direct it back toward the surface. In the absence of this natural greenhouse effect, Earth would be a frozen, and most likely lifeless, planet.

Another natural factor affecting Earth's climate—this one measured on timescales of several millennia—involves cyclical variations in the geometry of Earth's orbit around the sun. These variations alter the distribution of solar radiation over the surface of Earth and are responsible for the coming and going of the ice ages every one hundred thousand years or so. In addition, small variations in the brightness of the sun drive minor changes in Earth's surface temperature over decades and centuries. Explosive volcanic activity, such as the Mount Pinatubo eruption in the Philippines in 1991, also affects Earth's climate. These eruptions inject highly reflective particles called aerosol into the upper part of the atmosphere, known as the stratosphere, where they can reside for a year or longer. These particles reflect some of the incoming sunlight back into space and cool Earth's surface for years at a time.

## Human Progress Puts Pressure on Natural Climate Patterns

Since the dawn of the industrial revolution some two centuries ago, however, humans have become the principal drivers of climate change. The burning of fossil fuels—such as oil, coal, and natural gas—has led to an increase in atmospheric levels of carbon dioxide, a powerful greenhouse gas. And farming practices have led to increased atmospheric levels of methane, another potent greenhouse gas. If humanity continues such activities at the current rate through the end of this century, the concentrations of greenhouse gases in the atmosphere will be higher than they have been for tens of millions of years. It is the unprecedented rate at which we are amplifying the greenhouse effect, warming Earth's surface, and modifying our climate that causes scientists so much concern.

## The Role of Scientists in Climate Observation and Projection

Scientists study Earth's climate not just from observation but also from a theoretical perspective. Modern-day climate models successfully reproduce the key features of Earth's climate, including the variations in wind patterns around the globe, the major ocean current systems such as the Gulf Stream, and the seasonal changes in temperature and rainfall associated with Earth's annual revolution around the sun. The models also reproduce some of the more complex natural oscillations of the climate system. Just as the atmosphere displays random day-to-day variability that we term "weather," the climate system produces its own random variations, on timescales of years. One important example is the phenomenon called El Niño, a periodic warming of the eastern tropical Pacific Ocean surface that influences seasonal patterns of temperature and rainfall around the globe. The ability to use models to reproduce the climate's complicated natural oscillatory behavior gives scientists increased confidence that these models are up to the task of mimicking the climate system's response to human impacts.

To that end, scientists have subjected climate models to a number of rigorous tests of their reliability. James Hansen of the NASA Goddard Institute for Space Studies performed a famous experiment back in 1988, when he subjected a climate model (one relatively primitive by modern standards) to possible future fossil fuel emissions scenarios. For the scenario that most closely matches actual emissions since then, the model's predicted course of global temperature increase shows an uncanny correspondence to the actual increase in temperature over the intervening two decades. When Mount Pinatubo erupted in the Philippines in 1991, Hansen performed another famous experiment. Before the volcanic aerosol had an opportunity to influence the climate (it takes several months to spread globally throughout the atmosphere), he took the same climate model and subjected it to the estimated atmospheric aerosol distribution. Over the next two years, actual global average surface temperatures proceeded to cool a little less than 1°C (1.8°F), just as Hansen's model predicted they would.

Given that there is good reason to trust the models, scientists can use them to answer important questions about climate change. One such question weighs the human factors against the natural factors to determine responsibility for the dramatic changes currently taking place in our climate. When driven by natural factors alone, climate models do not reproduce the observed warming of the past century. Only when these models are also driven by human factors—primarily, the increase in greenhouse gas concentrations—do they reproduce the observed warming. Of course, the models are not used just to look at the past. To make projections of future climate change, climate scientists consider various possible scenarios or pathways of future human activity.

The earth has warmed roughly 1°C since preindustrial times. In the "business as usual" scenario, where we continue the current course of burning fossil fuel through the twenty-first century, models predict an additional warming anywhere from roughly 2°C to 5°C (3.6°F to 9°F). The models also show that even if we were to stop fossil fuel burning today, we are probably committed to as much as 0.6°C additional warming because of the inertia of the climate system. This inertia ensures warming for a century to come, simply due to our greenhouse gas emissions thus far. This committed warming introduces a profound procrastination penalty for not taking immediate action. If we are to avert an additional warming of 1°C, which would bring the net warming to 2°C—often considered an appropriate threshold for defining dangerous human impact on our climate—we have to act almost immediately.

## Long-Term Warming May Bring About Extreme Changes Worldwide

In the "business as usual" emissions scenario, climate change will have an array of substantial impacts on our society and the environment by the end of this century. Patterns of rainfall and drought are projected to shift in such a way that some regions currently stressed for water resources, such as the desert southwest of the United States and the Middle East, are likely to become drier. More intense rainfall events in other regions, such as Europe and the midwestern United States, could lead to increased flooding. Heat waves like the one in Europe in summer 2003, which killed more than thirty thousand people, are projected to become far more common. Atlantic hurricanes are likely to reach greater intensities, potentially doing far more damage to coastal infrastructure.

Furthermore, regions such as the Arctic are expected to warm faster than the rest of the globe. Disappearing Arctic sea ice already threatens wildlife, including polar bears and walruses. Given another 2°C warming (3.6°F), a substantial portion of the

Greenland ice sheet is likely to melt. This event, combined with other factors, could lead to more than 1 meter (about 3 feet) of sea-level rise by the end of the century. Such a rise in sea level would threaten many American East Coast and Gulf Coast cities, as well as low-lying coastal regions and islands around the world. Food production in tropical regions, already insufficient to meet the needs of some populations, will probably decrease with future warming. The incidence of infectious disease is expected to increase in higher elevations and in latitudes with warming temperatures. In short, the impacts of future climate change are likely to have a devastating impact on society and our environment in the absence of intervention.

## Strategies for Confronting Climate Change

Options for dealing with the threats of climate change include both adaptation to inevitable changes and mitigation, or lessening, of those changes that we can still affect. One possible adaptation would be to adjust our agricultural practices to the changing regional patterns of temperature and rainfall. Another would be to build coastal defenses against the inundation from sea-level rise. Only mitigation, however, can prevent the most threatening changes. One means of mitigation that has been given much recent attention is geoengineering. This method involves perturbing the climate system in such a way as to partly or fully offset the warming impact of rising greenhouse gas concentrations. One geoengineering approach involves periodically shooting aerosol particles, similar to ones produced by volcanic eruptions, into the stratosphere—essentially emulating the cooling impact of a major volcanic eruption on an ongoing basis. As with nearly all geoengineering proposals, there are potential perils with this scheme, including an increased tendency for continental drought and the acceleration of stratospheric ozone depletion.

The only foolproof strategy for climate change mitigation is the decrease of greenhouse gas emissions. If we are to avert a dangerous 2°C increase relative to preindustrial times, we will probably need to bring greenhouse gas emissions to a peak within the coming years and reduce them well below current levels within the coming decades. Any strategy for such a reduction of emissions must be international and multipronged, involving greater conservation of energy resources; a shift toward alternative, carbon-free sources of energy; and a coordinated set of governmental policies that encourage responsible corporate and individual practices. Some contrarian voices argue that we cannot afford to take such steps. Actually, given the procrastination penalty of not acting on the climate change problem, what we truly cannot afford is to delay action.

Evidently, the problem of climate change crosses multiple disciplinary boundaries and involves the physical, biological, and social sciences. As an issue facing all of civilization, climate change demands political, economic, and ethical considerations. With the Confronting Global Warming series, Greenhaven Press addresses all of these considerations in an accessible format. In ten thorough volumes, the series covers the full range of climate change impacts (water and ice; extreme weather; population, resources, and conflict; nature and wildlife; farming and food supply; health and disease) and the various essential components of any solution to the climate change problem (energy production and alternative energy; the role of government; the role of industry; and the role of the individual). It is my hope and expectation that this series will become a useful resource for anyone who is curious about not only the nature of the problem but also about what we can do to solve it.

*Michael E. Mann*

*Michael E. Mann is a professor in the Department of Meteorology at Penn State University and director of the Penn State Earth System*

*Science Center. In 2002 he was selected as one of the fifty lead-ing visionaries in science and technology by* Scientific American. *He was a lead author for the "Observed Climate Variability and Change" chapter of the Intergovernmental Panel on Climate Change (IPCC) Third Scientific Assessment Report, and in 2007 he shared the Nobel Peace Prize with other IPCC authors. He is the author of more than 120 peer-reviewed publications, and he recently coauthored the book* Dire Predictions: Understanding Global Warming *with colleague Lee Kump. Mann is also a co-founder and avid contributor to the award-winning science Web site RealClimate.org.*

# CHAPTER 1

# Population and the Human Imprint on Earth: An Introduction

The history of humankind is inextricably linked to the land, the atmosphere, the ocean, and the natural resources of the earth. Early in human history, the earth, its climate, and its resources exerted a tremendous impact on humans, while humans generally had a small impact on the earth's ecosystem. As civilization has advanced and population has increased, however, humans' impact on the earth has become more and more significant.

## The Earth's Ecosystem and Human Evolution

Modern humans' earliest ancestors were hunters and gatherers. They existed by foraging for edible plants and hunting animals in the wild. Neither plants nor animals were cultivated or domesticated. Humans were nomads and opportunists, hunting animals and collecting plants that were nearby and then moving to find new food sources when necessary. The size of these hunting and gathering societies was small, and their organization was minimal. Because their nomadic way of life prevented the accumulation of possessions, they were generally egalitarian—there were no "haves" or "have nots." Because early humans moved around they generally did not exploit any single plant or animal too heavily. Instead they used many resources in many areas lightly, and their impact on the earth was small. It is generally believed

## Was the Mayan Population "Collapse" Caused by Climate?

Between A.D. 750 and 1150, the Classic Mayan civilization of southern Mexico and Central America underwent a dramatic transformation involving complex changes in Mayan society and an apparent collapse of population size by 70 percent or more. Archaeologists have long argued about the root causes of this collapse, and many explanations have been proposed for this enigmatic story. Could an understanding of the earth system context help unravel the causes and effects involved in the population collapse and the major transformations that occurred in Mayan civilization during this time? Over the past fifteen years, evidence has been accumulated from sediment cores taken from lakes in the region that may help illuminate this relationship. These detailed sedimentary records show that the climate history over the period of collapse consisted of a series of protracted droughts, separated by intervening moister periods. The timing of these droughts coincides with indications from geological records of dry conditions elsewhere in the tropical Americas. Although many scientists have argued for a linkage between this history of drought and the archaeological record of declining Mayan population size, the connection remains controversial.

SOURCE: **National Research Council**, *Understanding Climate's Influence on Human Evolution*. **Washington, DC: The National Academies Press, 2010.**

that the population of humans during these early years was never more than 10 million at any one time.

Whereas humans' impact on the earth was minimal during that early time, the earth's impact on humans was dramatic. Evidence suggests that the planet's climate may have affected human evolution and led to the diversity within the human race that exists today. Data from ice cores show that early humans lived through a thousand-year ice age about seventy-one thousand years ago—a time that coincides with a bottleneck in human evolution (a significant reduction in the size and genetic makeup of the population). An ice age is an extended period of very cold temperatures during which much of the earth is covered with sheets of ice. Scientists think this ice age may have been caused by a volcanic eruption in Indonesia of Mount Toba, which spewed huge amounts of volcanic ash into the atmosphere. Genetic evidence suggests that this ice age devastated the human population, reducing it to a mere ten thousand adults on the entire planet. It is hypothesized that these ten thousand survivors found refuge from the difficult climate and settled in isolated pockets in the tropics, mainly in equatorial Africa. According to anthropologist Stanley H. Ambrose from the University of Illinois, genetic evidence suggests that the various racial groups of modern man arose from each of these isolated population groups.[1]

The transition of humans from hunting and gathering societies to an agrarian way of life may also have been shaped by the earth's climate. Ofer Bar-Yosef, an Israeli archeologist and professor at Harvard University, thinks a cooling trend caused a decline in food availability, which caused our ancestors to begin sowing and harvesting wheat and other cereal crops. Bar-Yosef's hypothesis is based on the archeological remains of an ancient Middle Eastern culture—the Natufian culture. Sometime between approximately 15,000 and 11,000 years ago, the Natufians occupied the area around the ancient city of Jericho. According to Bar-Yosef, the Natufians first domesticated wheat and rye about 11,500 years ago during a period in the earth's climate known as the Younger Dryas. The Younger Dryas was a cool, dry episode during an otherwise warm period in the earth's climate. Bar-Yosef contends that the cool, dry conditions of the Younger Dryas made the wild relatives of wheat and barley harder to find. According to Bar-Yosef, "The beginning of cultivation emerged from an environment of stress that forced people to rely more heavily on cultivated species. The Natufians, seeing the depletion of natural fields, came to the conclusion that they should start planting them instead of harvesting in the wild."[2]

## The Shift to Agriculture Intensifies Human Impact

Ancient farmers, like the Natufians, had more of an impact on the earth than their hunting and gathering forebears. Living in one spot permanently meant exploiting a relatively small amount of land very intensively (rather than exploiting a large amount of land lightly, as hunter-gatherers did), and over a long period. Planting crops typically requires a modification of the land and the nearby environment, particularly in regions like the Middle East, where water can be scarce. Archeological excavations of ancient agricultural sites in the Middle East show that ancient farmers dug up the ground to build channels to harvest the run-off from mountain slopes. They also built terraced wadis (river valleys) to capture floodwaters when it rained. Agricultural societies' consumption of large amounts of water for crops had, and still has, a significant impact on the world's ecosystem.

The transition from hunting and gathering societies to agricultural societies is called the first agricultural revolution, or the Neolithic Revolution (referring to the Neolithic period of history, also called the New Stone Age). Agriculture-based societies built permanent dwellings and accumulated a surplus of food and goods. These societies could support more people than hunting and gathering societies, and the rate of reproduction increased. Surplus food made possible an increasing division of labor; the development of social and political structures; and the creation of villages, towns, and eventually cities. Because of the success of agricultural societies, the human population on the

---

*Following pages: A neolithic agricultural settlement once stood at Skara Brae in Scotland. Human settlements have had an increased impact on the earth since the New Stone Age.* Patrick Dieudonne.

earth went from fewer than 10 million to 50 million in a few thousand years.

As time passed and civilization progressed from the Neolithic Age through the Bronze and Iron ages, humans continued to exert a greater influence on the earth. Agriculture, though still fairly primitive, increased in size and scale and permanently changed the landscape of the earth. Vast areas of woodland were cleared to make way for crops, and over time the soil in many regions became increasingly acidic and uncultivable. In his book *The Human Impact on the Natural Environment*, British geography professor Andrew Goudie called humans' impact on Earth during this time "an assault on the landscape."[3] During this period, the population of the earth showed a gradual increase each century. During the time of Jesus, the earth's population is estimated to have been around 200 million people.

The number of people on the planet waxed and waned from that time through the Middle Ages. The concentration of people in cities caused problems such as the buildup of waste and increased rates of poverty and disease, all of which led to widespread mortality. For instance, the Bubonic Plague, or Black Death, is thought to have reduced the world's population—mostly in Europe—from an estimated 450 million in A.D. 1340 to between 350 million and 375 million by the beginning of the fifteenth century. Additionally, wars such as the Crusades (approximately 1095–1291) caused the deaths of millions of people.

By the middle of the seventeenth century the human population reached about 500 million and began an upward climb that has yet to recede. Wars and poverty were not eradicated—humanity would suffer through hungry times and many devastating conflicts. But the population losses from such circumstances were more than offset by scientific and technological advances, particularly ones in medicine, which controlled disease and prolonged life. Such progress propelled the human population to 813 million by the beginning of the nineteenth century and to 1.55 billion by the beginning of the twentieth century.

## The Twentieth Century

The human population skyrocketed in the twentieth century. It took more than ten thousand years for the earth's population to reach the 1 billion mark. In the twentieth century alone, however, the population rose by nearly 5 billion people. As of March 16, 2010, the US Census Bureau figured the world's population at 6.8 billion people. The United Nations estimates that by 2050 the earth will be home to more than 9 billion people.

The industrial revolution, which began in England in the mid-eighteenth century, evolved significantly in the twentieth century. Like the earlier agricultural revolution, the industrial revolution changed human society in dramatic ways. It involved a shift from small-scale production of goods by hand to large-scale production of goods by machines. The twentieth century brought the harnessing of electricity, and numerous appliances helped humans work more efficiently and with less difficulty. It also brought the advent of the automobile. The industrialization of society gave rise to sweeping increases in production capacity that affected all basic human needs, including food production, medicine, housing, and clothing.

*In the twentieth century, the changing climate revealed the power humans could have over the planet.*

Industrialization also significantly increased humans' impact on the earth. Gaseous emissions from the fossil fuels that drove the industrial revolution—coal, oil, and natural gas—entered the atmosphere and harmed human health, wildlife, and the environment. Sulfur dioxide ($SO_2$) and mercury emitted from the burning of coal for electricity production caused acid rain or accumulated in wildlife. Toxic substances, such as nitrogen oxides

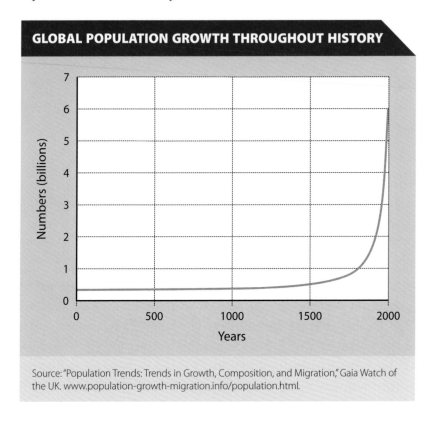

**GLOBAL POPULATION GROWTH THROUGHOUT HISTORY**

Source: "Population Trends: Trends in Growth, Composition, and Migration," Gaia Watch of the UK. www.population-growth-migration.info/population.html.

(NO$_x$), particulate matter, and volatile organic compounds in tailpipe emissions from automobiles were causing smog and exacerbating diseases including asthma, cancer, respiratory illness, and heart disease. It also became apparent that carbon dioxide (CO$_2$)—a seemingly harmless gas that is expelled with every human and animal exhalation and is the final by-product of all fossil-fuel combustion—was affecting the very climate itself.

As the human population has increased in size and become more advanced, our imprint on the earth has gotten larger and larger. In his 2000 book, *Something New Under the Sun*, Georgetown University environmental historian J.R. McNeill wrote, "Our genus, Homo, has altered earthly environments throughout our career, about 4 million years. But there has never been anything like the twentieth century."[4] In the twentieth century, the changing climate revealed the power humans could have over the planet. And, in turn, as governments and societies scramble to prepare for the impact of human-induced global warming, the power the climate still has over humanity becomes startlingly clear.

## Notes

1. Stanley H. Ambrose, "Late Pleistocene Human Population Bottlenecks, Volcanic Winter, and Differentiation of Modern Humans," *Journal of Human Evolution*, 1998, pp. 623–651.
2. Ofer Bar-Yosef, Clark Erickson, and T. Douglas Price, "First Farmers," *The Why Files*, 2000. http://whyfiles.org.
3. Andrew Goudie, *The Human Impact on the Natural Environment*. Cambridge, MA: MIT Press, 2000, p. 175.
4. J.R. McNeill, *Something New Under the Sun*. New York: Norton, 2000.

# CHAPTER 2

# The Link Between Population Dynamics and Resources

Each year that the world population grows, humanity places more demands on the earth's resources. Since the late 1960s, the world's population has grown by about 70 million to 88 million people each year. Between 2009 and 2010, the population increased by about 74 million to reach the 6.8 billion mark. Although experts predict the rate of population growth to decline steadily, it is estimated that by 2050 the earth will be home to 9 billion people. As the world's population increases, our need for the earth's resources—food, water, land, and energy—also increases, as does our production of wastes and pollution. The magnitude of humanity's footprint on the planet ultimately depends on the numbers and characteristics of the earth's population and the efficiency of our use of the planet's resources.

## Theories About the Connection Between Population and Resources

Throughout history, scientists have proposed many theories to describe the relationship between the human population and the earth's resources. One of the most famous of these theories was provided by British scholar Thomas Robert Malthus in the late eighteenth century. During this time the human population was creeping toward the 1 billion mark. Malthus maintained that the population would double every twenty-five years if left unchecked. He also believed that the earth's resources would never be able to keep pace with this kind of exponential growth. In a 1798 treatise, *An Essay on the Principle of Population*, Malthus writes, "the power of population is indefinitely greater than the power in the earth to produce subsistence for man."[1] Malthus predicted that the population would grow until the earth's resources were no longer adequate to sustain humanity. Then, he predicted, there would be widespread poverty, starvation, and misery.

Malthus's theory of population doubling every twenty-five years did not hold true. His belief that unchecked population growth would be catastrophic for humanity is embraced by many modern scientists, however, such as Stanford University professor Paul Ehrlich, as well as Stanford's John Holdren, who was appointed science advisor to President Barack Obama. In a 1969 article published in the journal *BioScience*, the two professors argue, "if population control measures are not initiated immediately, and effectively, all the technology man can bring to bear will not fend off the misery to come."[2]

In 1971, Ehrlich and Holdren proposed a simple formula to measure the impact of human activities on the earth's environ-

*Contrary to Malthus's predictions, China's increasing population has also experienced an overall rise in the standard of living in the country.* AP/Zhao Xiaoming.

ment. The formula, I = P × A × T, maintains that humanity's environmental impact (I) on the earth equals the mathematical product of size of the *population* (P), level of consumption per person, or *affluence* (A), and amount of resources needed or wastes created by the *technology* used to make useful goods (T).[3] Although the IPAT formula may be an oversimplification, most scientists believe the impact to the earth from human activities is based upon some contribution of population, economic growth, and technology.

Not everyone believes population growth is necessarily bad. Some scientists think that having more human beings on the earth is a good thing. The late Julian Simon, an economist and author affiliated with the University of Maryland, was a well-known proponent of this belief. Simon contended that humanity has benefited as the population has grown larger. In a 1981 book titled *Ultimate Resource*, Simon writes, "the standard of living has risen along with the size of the world's population since the beginning of recorded time. There is no convincing economic reason . . . these trends toward a better life should not continue." Simon believed that humans are the earth's "ultimate resource" and that people are "not just more mouths to feed, but are productive and inventive minds that help find creative solutions to [humanity's] problems, thus leaving us better off over the long run."[4] Simon railed against the Malthusian premise that dwindling resources and increasing pollution, starvation, and misery were inevitable unless population growth is curbed.

## Population Changes: Birth and Death Rates

Irrespective of whether population growth is deemed "good" or "bad," its changes over time are important components of past world history and useful predictors for the future state of the world. Demographers, scientists who study populations, generally look at population trends in individual countries. They look particularly at birth and death rates, as these factors are most important in

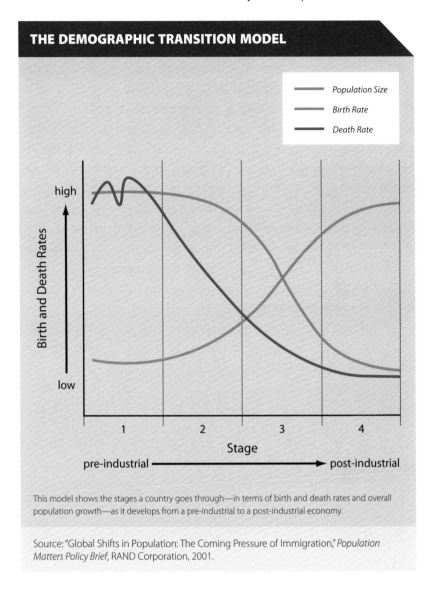

## THE DEMOGRAPHIC TRANSITION MODEL

This model shows the stages a country goes through—in terms of birth and death rates and overall population growth—as it develops from a pre-industrial to a post-industrial economy.

Source: "Global Shifts in Population: The Coming Pressure of Immigration," *Population Matters Policy Brief*, RAND Corporation, 2001.

determining whether a country's inhabitants will increase or decrease, and at what rate. Most industrialized nations have followed a similar pattern of population growth—affected by fluctuating birth and death rates over time—and this pattern is demonstrated by what is known as the demographic transition model.

At the time Malthus was devising his theory, differences in birth and death rates between industrialized and nonindustrialized countries were beginning to emerge. Previously in history, regions worldwide underwent high rates of birth and death, which resulted in balanced or very slow population growth. Malthus's lifetime (1766–1834) coincided with the beginning of the industrial revolution, which started in Great Britain and spread to countries in Europe and to the United States and Canada. The death rates in these countries declined because industrialization and medical advancements led to improvements in sanitation, food handling, and general personal hygiene. At the same time the birth rates in these countries remained high. As a result, by the middle of the nineteenth century most industrialized countries, including the United States, Canada, and Japan, experienced rapid population growth.

While populations were expanding in industrialized or developed countries, growth in nonindustrialized or developing countries—such as China, India, and many countries in Africa and South America—remained slow. The death rate in these countries did not start falling until well into the twentieth century. Once it started falling, however, it dropped rapidly. That circumstance, combined with a surging fertility rate, meant that the population of the developing world boomed in the twentieth century.

By that time, however, industrialized countries were undergoing another demographic change: declining fertility rates. Increasing female literacy, greater employment opportunities, and changing social structures caused women to have fewer children, delay childbirth, or forgo the experience altogether. The widespread availability of birth control beginning in the 1960s further intensified the decrease in fertility rates in these countries. The lower birth rate caused population growth in many countries to slow considerably. Several industrialized countries, including Italy, Japan, and Poland, have even seen population declines as birth rates have not kept up with death rates.

The birth rate in developing countries also started declining in the twentieth century, although it still outpaced the rate in developed countries. According to the United Nations, between 1995 and 2000, the birth rate in developed countries was 11 births per 1,000 people, while in less developed countries it was 23 births per 1,000 people. The 49 countries the United Nations considers the least developed—primarily in Africa and Asia— had birth rates of 39 per 1,000 people. The high birth rates in less developed countries are attributed primarily to the status of women in these countries. Few women there attend school or receive an education. In some countries women spend most of their time fetching water and caring for children. Family planning and contraception are not widespread.

*According to the UN, . . . 99 percent of the expected population growth [throughout the twenty-first century] will occur in less developed regions of the world.*

The difference between developed and developing countries was not as striking when comparing death rates. From 1995 to 2000, death rates in developed regions were 10 people per 1,000, compared with 14 per 1,000 in the least developed countries.

Although birth rates are falling and death rates everywhere are relatively low, the United Nations still projects that the world population could reach 9 billion by 2050. According to the UN, the birth rate in less developed regions will continue to fall throughout the twenty-first century. The world's population will still grow considerably, however, and 99 percent of the expected growth will occur in less developed regions of the world.

## Resource Consumption and Environmental Impact

The growing population of the world places more and more demands on land, water, energy, and other resources. Environmental scientist Jonathan Foley believes our use of the land—particularly for agriculture—is one of the most pressing environmental issues of our time. Writing in the journal *Yale Environment 360* in October 2009, Foley states "already, we have cleared or converted more than 35 percent of the earth's ice-free land surface for agriculture, whether for croplands, pastures or rangelands. In fact, the area used for agriculture is nearly 60 times larger than the area of all of the world's cities and suburbs. Since the last ice age, nothing has been more disruptive to the planet's ecosystems than agriculture."[5] Foley believes that population growth, rising meat and dairy consumption, and increasing needs for bioenergy sources will require a doubling or tripling of the world's agricultural production in the next thirty to forty years.

Although two thirds of the earth is covered in water, the vast majority of it is saltwater and therefore useless for most human needs. Only 3 percent of the earth's water is considered freshwater, available for drinking or agriculture. Of this amount, about 69 percent is contained in ice caps and glaciers, and approximately 31 percent is contained in deep underground aquifers. Only 0.3 percent of the earth's freshwater is contained in lakes, rivers, and streams, which are distributed unevenly throughout the world. For instance, freshwater is generally abundant in South America and Asia but scarce in North Africa and the Middle East. Population growth in dry areas of the world places greater demands on already tight supplies of freshwater. According to the nonprofit organization CARE, it is estimated that more than 40 percent of the world's population, or 2.7 billion people—mostly living in the developing world—will experience severe water scarcity by 2025.

As the world's population grows, so too does the need for fuels in cooking, heating, transportation, and industry. In many developing countries, people do not have access to the electric grid or modern fuels. They rely on wood, charcoal, straw, or dung for cooking and on car batteries or diesel generators for electricity. The use of wood for cooking and heat in developing countries contributes to the removal of forest land and the loss of habitat for wildlife. As countries become more populous and more industrialized, their consumption of energy, particularly from fossil fuels (coal, oil, and natural gas), generally increases significantly. In 1980, for instance, India consumed 4 quadrillion BTUs (British Thermal Units) of energy and China consumed 17 quadrillion BTUs. In 2006 both countries' energy consumption more than quadrupled as India consumed 18 quadrillion BTUs of energy and China consumed 74 quadrillion BTUs of energy. During this same time, the United States' total energy consumption increased from 78 quadrillion BTUs to 100 quadrillion BTUs.[6] Most of this energy use is generated by fossil fuels, which are in limited supply and are nonrenewable. As more countries become industrialized, the demands on the world's energy supplies increase.

In addition to causing an increase in energy consumption, industrialization also generates more waste and pollution, which stress and potentially damage the earth, its ecosystems, and its climate. The material goods of industrialized societies are packaged in plastics and other materials that ultimately are discarded in landfills across the world. Toxic wastes from manufacturing processes and fertilizers from industrialized agricultural operations often leech into water sources and contaminate them. Transportation and electricity generation, as well as heating of homes and businesses, release mercury, sulfur dioxide ($SO_2$), carbon dioxide ($CO_2$), and other such emissions into the air.

Carbon dioxide emissions, the biggest contributor to global warming, have risen in countries with increasing populations and industrialization. According to reported energy statistics, coal production and use in China has increased tenfold since the 1960s. As a result, China's industrial emissions of $CO_2$ have grown significantly. In 1950, China stood tenth among nations based on annual fossil-fuel $CO_2$ emissions. In 2007, however, China emitted 1.78 billion metric tons of carbon to become the world's largest emitter of $CO_2$. Beginning in 1950, India's $CO_2$

emissions began a dramatic upward climb, increasing by 5.7 percent each year. In 2007, India was ranked third in the world based on its level of fossil-fuel-related $CO_2$ emissions, while the United States was ranked second. US fossil-fuel emissions have doubled since the 1950s, but the US share of global emissions has declined from 44 percent to 20 percent in the same interval because of higher growth rates in other countries.[7]

In 2010, Australian researchers ranked the impact of each country on the earth. The researchers looked at a country's impact on its own resources, as well as its impact on the planet as a whole. The researchers examined forest loss, habitat conversion, the state of fisheries, fertilizer use, water pollution, species threats, and carbon emissions from land use. In regard to their impact on their own resources, or their proportional impact, the world's ten worst environmental performers were Singapore, Korea, Qatar, Kuwait, Japan, Thailand, Bahrain, Malaysia, the Philippines and the Netherlands. In terms of their impact on the entire planet, or their absolute impact, the ten worst performers were Brazil, the United States, China, Indonesia, Japan, Mexico, India, Russia, Australia, and Peru. The researchers found that increasing wealth was the most important driver of environmental impact. Countries with greater total wealth had worse environmental records than poorer countries. In addition, the researchers found that "countries with higher total human populations and densities had greater proportional environmental impact, and those with lower population growth rates had a slightly lower proportional environmental impact."[8] Densely populated and highly urbanized countries had higher proportional natural forest loss, greater releases of carbon dioxide, and a higher proportion of species threatened with extinction because of habitat loss and pollution.

## Carrying Capacity

The concept of carrying capacity refers to the level of population growth of a particular species that can be sustained by an ecosystem. Biologists use the concept of carrying capacity to determine how much habitat must be conserved to maintain healthy wildlife populations. At carrying capacity, the species' population will have an impact on the resources of the ecosystem, but not to the point where the ecosystem can no longer sustain the population. Population growth beyond the carrying capacity will result in a population crash, however, as the ecosystem's resources will be depleted and will no longer be able to provide for all the individuals of a species.

Applying carrying capacity to the human population is more complex than applying it to other species. Aside from the biophysical limitations of the earth's natural resources—the available food, water, and energy supplies—there is a social aspect to the carrying capacity for humans. The social aspect accounts for quality of life issues, or standard of living. Americans have one of the world's highest standards of living. Although there are many people who live in poverty in the United States, on average Americans have relatively small families, large homes, many possessions, plentiful food, and clean water. These qualities are absent in most of the developing world, where homes are smaller, families are bigger, food is not plentiful, and water is scarce. Sustaining the human population at the average American's standard of living necessarily requires more resources than sustaining the human population at a lower standard of living.

Ecologists, economists, and other scientists have tried to estimate the carrying capacity of the earth. In Joel Cohen's book *How Many People Can the Earth Support?* (1995) the author summarizes estimates that scientists have made going back almost to Malthus's time. The estimates vary dramatically from 0.5 billion to 14 billion, depending on the method used and the standard of living and technological improvements that are assumed. According to some of these estimates, humans have already exceeded the carrying capacity of the earth. In his book, Cohen never comes to a conclusion on the earth's carrying capacity. He does think that human population growth will come

# The Case for a New Human-Induced Epoch

In 2010 four notable scientists provided evidence that human beings had changed the earth so much that an entirely new epoch in geological time had been created. Geologists Jan Zalasiewicz and Mark Williams from the University of Leicester in the United Kingdom; Will Steffen, director of the Australian National University's Climate Change Institute; and Paul Crutzen, a Nobel-winning atmospheric chemist from the Max Planck Institute in Germany, published their evidence for a new epoch in the February 25, 2010, issue of *Environmental Science & Technology*. Crutzen, best known for his research on ozone depletion, had proposed the designation of a new epoch ten years earlier, calling it the Anthropocene to denote the idea that it is dominated by human activity.

How have the activities of humans altered the earth so significantly, Crutzen and his fellow scientists asked? The answers, they write,

> boil down to the unprecedented rise in human numbers since the early nineteenth century—from under a billion then to over six billion now, set to be nine billion or more by midcentury. This population growth is intimately linked with massive expansion in the use of fossil fuels, which powered the Industrial Revolu-

to an end at some point, however, probably in the twenty-first century.[9]

Does the earth have a carrying capacity for humans, or will humanity's technological ingenuity overcome the planet's inherent resource limitations? The answer to this question may become apparent sometime in the twenty-first century.

## Notes

1. John Malthus, *An Essay on the Principle of Population*. New York: Penguin Classics, 1798.
2. Paul Ehrlich and John Holdren, "Population and Panaceas: A Technological Perspective," *Bioscience*, December,1969.

tion, and allowed the mechanization of agriculture that enabled those additional billions to be fed.

The presence of megacities, said the scientists, is the most visible physical effect of this human expansion on the earth. The most profound effects, however, are chemical changes to a group of trace components in the earth's atmosphere, known collectively as greenhouse gases. According to the scientists, changes in the atmospheric levels of greenhouse gases are increasing the earth's surface temperature with far-lasting consequences.

The most dire of these consequences, write the scientists, is a dramatic loss of the earth's species. This species loss is being instigated by climate change and other human stressors, such as fragmented habitat, invasive species, and predation. They predict that this wave of species extinctions will be the sixth-greatest extinction event in the known history of the earth.

Zalasiewicz, Williams, Steffen, and Crutzen acknowledge that it will be difficult and contentious to get a formal designation for a new epoch. "However these debates unfold," they state, "the Anthropocene represents a new phase in the history of both humankind and of the earth, when natural forces and human forces became intertwined, so that the fate of one determines the fate of the other. Geologically, this is a remarkable episode in the history of this planet."

3. Paul Ehrlich and John Holdren, "Impact of Population Growth," *Science*, March 26, 1971.
4. Julian Simon, *Ultimate Resource*. Princeton, NJ: Princeton University Press, 1981.
5. Jonathan Foley, "The Other Inconvenient Truth: The Crisis in Global Land Use," *Yale Environment 360*, October 5, 2009.
6. Energy Information Administration, *International Energy Annual 2006*, December 19, 2008.
7. G. Marland, T.A. Boden, and R.J. Andres, "Global, Regional, and National Fossil-Fuel $CO_2$ Emissions," Carbon Dioxide Information Analysis Center, 2010.
8. Corey Bradshaw, Xingli Giam, and Navjot Sodhi, "Evaluating the Relative Environmental Impact of Countries," *PLoS ONE*, 2010.
9. Joel Cohen, *How Many People Can the Earth Support?* New York: Norton, 1995.

# CHAPTER 3

# Population's Impact on Climate Change

Global warming is a modern phenomenon directly related to industrialization and the growth of the human population. Since the nineteenth century, humans have extracted carbon-rich fossil fuels from deep within the earth. The burning of fossil fuels to release energy—and provide heat or generate electricity—spurred a revolution that allowed humankind to prosper and advance. The burning of such fuels has also released enough carbon dioxide into the air to change the climate. As long as carbon-rich fuels drive the world's economic engine, or until humanity finds efficient methods to prevent carbon dioxide from entering the atmosphere, population growth and economic development will pose the challenges of climate change.

## Population Growth

Simply put, as the human population grows, so too do emissions of carbon dioxide ($CO_2$), the most important greenhouse gas. When global $CO_2$ emissions are averaged over the world population, each person is responsible for the emission of about four metric tons of $CO_2$ per year.[1] Statistics show that this amount has stayed relatively constant since at least 1980. This means that as each new person is added to the world's population rolls, an additional four metric tons of $CO_2$ are emitted into the atmosphere each year of that person's life.

For this reason, many people think government policies that lower birth rates, particularly in developing countries where most population growth is occurring, can help mitigate climate change. Writing in the *Bulletin of the Atomic Scientists* in October 2009, Laurie Mazur, director of the Population Justice Project, states that slowing population growth is a "sensible, low-tech way to reduce our collective carbon footprint for relatively little cost."[2]

Mazur distinguishes population growth from population control. According to Mazur, population control programs, such as China's strict one-child policy, are coercive and violate individual human rights. Population growth programs, by contrast, seek to slow growth by ensuring that individuals have access to contraception and reproductive-health services so they can make voluntary choices about childbearing. These types of programs also empower girls and women through education and employment opportunities. Ecologist Frederick Meyerson holds similar views. Writing in the *Bulletin of the Atomic Scientists* in February 2008, Meyerson asserts that about two hundred million women in developing countries would like to delay or prevent pregnancy, but they lack access to effective contraception. "Reaching and helping these women and their partners is critical for climate and human development policy," states Meyerson.[3]

## Consumerism and Consumption

Many people and organizations think that focusing on population growth is misguided because climate change is affected by much more than the sheer size of the human population. The Sierra Club, the Worldwatch Institute, and others believe excessive consumption driven by consumerism is the main culprit in climate change, as well as in most other environmental problems affecting the world today. The Worldwatch Institute defines consumerism as the cultural orientation that leads people to find meaning, contentment, and acceptance through what they consume. Consumerism causes people to want larger

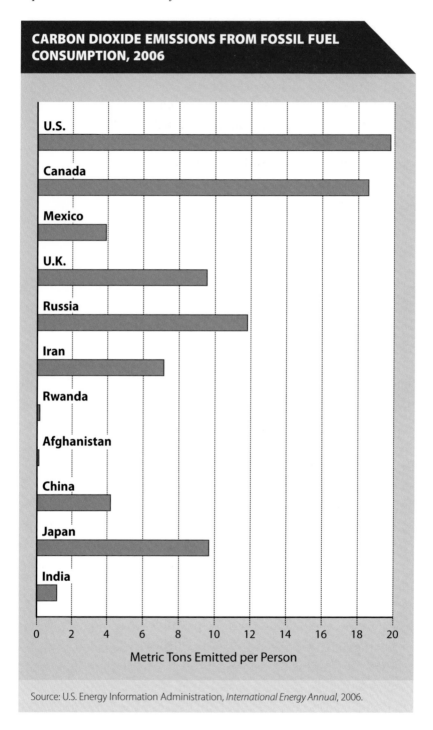

## CARBON DIOXIDE EMISSIONS FROM FOSSIL FUEL CONSUMPTION, 2006

- U.S.
- Canada
- Mexico
- U.K.
- Russia
- Iran
- Rwanda
- Afghanistan
- China
- Japan
- India

0  2  4  6  8  10  12  14  16  18  20

Metric Tons Emitted per Person

Source: U.S. Energy Information Administration, *International Energy Annual*, 2006.

homes, the biggest televisions, the most expensive cars, every electronic device on the market, and rich foods. To feed consumerism, more fossil fuels, minerals, and metals are being extracted from the earth, more trees are cut down, and more land is cleared. According to the Worldwatch Institute's *State of the World 2010*, "the exploitation of the Earth's resources to maintain ever higher levels of consumption has put increasing pressure on Earth's systems and in the process has dramatically disrupted the ecological systems on which humanity and countless other species depend."[4]

---

*As China's and India's economies have grown, so too have their carbon emissions. Between 1980 and 2001, China's $CO_2$ emissions grew by 111 percent. During the same period, India's $CO_2$ emissions more than tripled.*

---

The Worldwatch Institute and others state that excessive consumption is primarily a characteristic of rich industrialized nations. They point out that, while the average global per capita (per person) emissions of $CO_2$ have remained constant at about 4 metric tons annually, there is a huge variation between countries. For example, the United States has a bigger per person impact on climate change than most other countries. Per capita US emissions of $CO_2$ in 2006 were nearly twenty times larger than those of India or Africa and four to five times larger than those of China. According to a study by Princeton ecologist Stephen Pacala, the world's richest five hundred million people (roughly 7 percent of the world's population) are currently responsible for 50 percent of the world's carbon dioxide emissions, while the poorest three billion are responsible for just 6 percent.[5] In its *State of the World 2008*, the Worldwatch Institute states, "Clearly, Western nations have been the key driver of climate change so far. Between 1950 and 2000, the United States was responsible for 212 gigatons of carbon dioxide, whereas India was responsible for less than 10 percent as much. So it is clear that the richest people on the planet are appropriating more than their fair share of 'environmental space.'" The report goes on to state, "It is these countries that most urgently need to redirect their consumption patterns, as the planet cannot handle such high levels of consumption."[6]

Environmental groups worry that many developing countries are aspiring to attain the Western world's lifestyle and overconsumptive ways. This concern is heightened in regard to countries such as China and India, which have fast-growing populations and fast-growing economies. China and India have been recognized as the two largest and most booming developing economies in the world. To date, the two nations are home to more than one third of the world population and contribute 19.2 percent of world GDP (gross domestic product)—a measure of economic output. The economic successes in these developing countries have resulted in considerable improvements in quality of life and have many of those nations' citizens eager to obtain a lifestyle similar to that of Western societies. This desire is worrying to the United Nations Population Fund (UNFPA). According to the agency, "the current lifestyles and consumption patterns of the rich simply cannot be generalized to the world's entire population without causing severe environmental imbalances."[7]

As China's and India's economies have grown, so too have their carbon emissions. Between 1980 and 2001, China's $CO_2$ emissions grew by 111 percent. During the same period, India's $CO_2$ emissions more than tripled.[8] According to UNFPA, China and India are on par to contribute more than half of global $CO_2$ emissions by 2050. The Sierra Club asserts that, "Population, global warming and consumption patterns are inextricably linked in their collective global environmental impact. As developing countries' contribution to global emissions grows, population size and growth rates will become significant factors in magnifying the impacts of global warming."[9]

## Urbanization

In addition to population size, consumption levels, and economic growth, other demographic factors appear to be connected to $CO_2$ emissions. For instance, researchers have found that as a population ages, its economy slows and hence its emissions fall. Brian O'Neill, leader of the Population and Climate Change program at the National Center for Atmospheric Research in Boulder, Colorado, has been studying the effects of specific factors such as aging, household size, and urbanization on carbon emissions. He and his colleagues have found that smaller US households are associated with higher per-capita carbon-based energy expenditures (for fuel and utility purchases, for example) than larger ones, and that younger US households have higher per capita carbon-intensive expenditures than older ones. O'Neill's research also found that urbanization—the migration of people from rural areas to cities—is associated with increased carbon emissions.[10]

Since the latter half of the twentieth century, urbanization has been occurring at an unprecedented rate. According to the United Nations, in 1950, fewer than 30 percent of the world's people were urban dwellers. By 2010 more than half the world's almost 7 billion people lived in cities. The United Nations estimates that by 2050, more than two thirds of the world will live in urban areas, and most urban growth will be occurring in developing countries.

Cities are associated with climate change in numerous ways. First, people living in urban areas tend to be more affluent and have higher consumption levels. From the standpoint that consumption and affluence are associated with higher greenhouse gas emissions, so too are cities. Second, urban areas, with their concentration of asphalt and skyscrapers, can absorb heat and create heat islands. This causes people in cities to use more energy—and generate more greenhouse gas emissions—as they try to keep cool. Third, many cities are connected to urban sprawl—scattered development that necessitates an extensive infrastruc-

# The Impact of US Households on Climate Change

"Households," and the resources and energy Americans consume to support them, [are] important demographic [variables] in calculating population's climate change linkages. Every household has a minimum number of possessions, occupies a certain amount of space, and emits certain waste and/or pollutants. However, the extent of environmental stress, including climate change, that is linked to "households" depends on three main factors: household size (the number of people within a given household) and number of households; size of [home] (square footage of a house); and the amount of land surrounding and used to build [a home]. In recent decades, while the average US household size has decreased, the number of households has increased significantly, and the amount of "living space" in and around homes has risen. The number of people per household was 2.6 people in 2000, down from 3.1 in 1970 (or one fewer person for every two households), and as a result, the number of households has increased markedly. In 2000, there were nearly 14 million more housing units nationwide than a decade earlier. Between 1970 and 2000, average household size declined from 3.1 to 2.6 persons (one fewer person for every two households), resulting in demand for new housing units in addition to that already needed to keep pace with population growth. Smaller household size in the face of population growth is one reason behind a nationwide build-

ture, increases traffic, and consumes open space, all of which contribute to climate change.

An international study of ten cities found that some cities contribute more to climate change than others. In a paper published in *Environmental Science & Technology*, Christopher Kennedy and his colleagues analyze how climate, power generation, transportation, waste processing, and other factors contribute to a city's greenhouse gas emissions. Among the

ing boom: between 2000 and 2006, 10 million new housing units were built. In addition, the number of second homes is at [a historical peak]: second home ownership increased from 5,537,000 units to 6,489,000 units, a 17% increase or 2.7% annual increase from 1999 to 2005. The average size of new, single-family homes has expanded steadily, reaching more than 2,300 square feet by 2004. Nearly 40% of new single-family homes are over 2,400 square feet, double the proportion in 1987. With more people living in "super-sized" houses that occupy more land, the amount of resources (from lumber to plastic) used for new construction is rapidly on the rise, and more energy is consumed for heating and cooling (thus more fossil fuels burned). With many houses built in new sprawling communities—rather than clustered or city-based development that often uses existing "footprints" to build new homes—development begins from scratch . . . there is higher resource use, and fragmentation of open space. Data shows that increases in the average [lot] sizes, on which new houses are built, are prevalent in many suburban areas. About 55% of farmland developed since 1994 has gone to houses built on lots ten acres or larger. More than 3,000 square miles of land is converted annually to residential development over one acre in size. Energy use within each household is also a key population-climate change link. The US residential sector is the largest of that [energy-use] sector worldwide, and household appliances are the fastest growing energy consumers nationwide.

SOURCE: Victoria Markham, "U.S. Population, Energy & Climate Change," Center for Environment & Population, 2008.

ten cities studied, Denver had the highest overall greenhouse gas emissions, with levels two to five times higher than those of other cities. The researchers found that its heightened levels were due partly to its high use of electricity, heating and industrial fuels, and ground transportation. Los Angeles was second on the list, followed by Toronto and Cape Town (tied for third), Bangkok, New York city, London, Prague, Geneva, and Barcelona.[11]

## Poverty and Land Use

Whereas affluence is associated with higher emissions, poverty can also exacerbate climate change. Poverty often forces people to make choices that are bad for the environment. For instance, many poor, rural people in tropical regions practice a technique of farming called slash and burn, which can lead to deforestation, or permanent removal of woodland. Slash and burn is an ancient form of agriculture where forests are burned down to provide land for agriculture. Once soil nutrients become depleted and the field becomes unproductive, it is abandoned and generally the farmers move on to burn and cultivate other forest land. Slash and burn is not inherently bad. Generally, native plant and tree species regenerate and reclaim the abandoned field. However, when practiced incorrectly and on too large a scale, however, slash and burn can cause permanent loss of woodland.

Slash and burn is contributing to deforestation in many tropical areas as poor people are flocking to the forest in search of food. Poverty and a lack of employment have caused large numbers of people to move to tropical forest areas seeking to grow enough food to feed their families. These subsistence farmers, as they are called, are practicing slash-and-burn agriculture and putting enormous pressure on forest land. Farmers are forced to clear new forests or return to old fields before the forest has adequately regenerated. This forest loss is an important component of climate change.

Tropical forests are very important in helping to mitigate climate change as they can absorb vast quantities of $CO_2$. When forests are permanently removed, climate change is exacerbated because the forest no longer absorbs $CO_2$ and because the carbon that was once stored in the forest typically ends up being released into the air through burning or decomposition. As a result, what was once a carbon sink becomes a carbon source. Deforestation

---

*Following page: Slash-and-burn agriculture has contributed to severe deforestation on the island of Madagascar.* Roberto Schmidt/AFP/Getty Images.

accounts for 12 percent to 20 percent of the human-caused greenhouse gas emissions in the world.

Poverty and slash-and-burn agriculture are not the only factors contributing to deforestation. Huge numbers of trees have been removed from tropical forest areas to satisfy the world's need for timber. Additionally, huge tracts of tropical forests are being cleared to become grazing land for livestock. According to a 2009 report by Greenpeace, Amazonian deforestation is being driven by world demand for beef and leather products. The report states that millions of tons of Brazilian beef and leather from cattle grazed on deforested land are exported to markets in the United States, Europe, Russia, and the Middle East. The report explains that the cattle sector in the Brazilian Amazon is the largest driver of deforestation in the world.[12]

## Talking About Population and Climate Change

It is clear that population and the consumption it begets are linked to climate change. Only recently have demography and population policies become part of climate change research, however, a circumstance attributed to the sensitive nature of talking about population in general. In the past, population policies have generally referred to population control programs, such as China's one-child policy, that have infringed on human rights. Additionally, according to Robert Engelman, author of *More: Population, Nature, and What Women Want*, it has been difficult to bring population into climate change discussions because most growth is occurring in poorer nations, whereas "most of the responsibility for human-induced global warming stems from the past behavior of wealthier nations, most of whose populations are now growing relatively slowly or not at all."[13]

Despite the political sensitivity associated with the issue, population is beginning to be considered in climate change discussions. Many population-oriented nonprofit organizations, such as the Population Reference Bureau and the Population

Council, are organizing seminars and discussions centering on climate change. Alternatively, ecologists and climate scientists are incorporating demographic data into their research. According to demographer and climate researcher Brian O'Neill, "the effect [population] could have on the quintessentially modern issue of climate change is something that's definitely worth looking into further. Demographers have a lot to say about it."[14]

## Notes

1. Energy Information Administration, "Table H.1cco2 World Per Capita Carbon Dioxide Emissions from the Consumption and Flaring of Fossil Fuels, 1980–2006," International Energy Annual 2006, December 2008.
2. Laurie Mazur, "A Neglected Climate Strategy: Empower Women, Slow Population Growth," *Bulletin of the Atomic Scientists*, October 12, 2009.
3. Frederick Meyerson, "Population and Climate Change," *Bulletin of the Atomic Scientists Web Edition Roundtable*, April 16, 2008.
4. Worldwatch Institute, *State of the World 2010: Transforming Cultures*. New York: W.W. Norton, 2010.
5. Worldwatch Institute, *State of the World 2010*.
6. Worldwatch Institute, *State of the World 2008: Innovations for a Sustainable Economy*, New York: W.W. Norton, 2008
7. UN Population Fund, "UNFPA Statement: Population and Climate Change," February 11, 2008. www.unfpa.org.
8. Worldwatch Institute, *State of the World 2010*.
9. Sierra Club, "Who Is Heating Up the Planet?" 2008.
10. In Sonia Schmanski, *Population and Climate Change: Relationships, Research, and Responses*, ed. Rachel Weisshaar. Woodrow Wilson International Center for Scholars, February 20, 2008. wilsoncenter.org,
11. Christopher Kennedy et al., "Greenhouse Gas Emissions from Global Cities," *Environmental Science & Technology*, October 1, 2009.
12. Greenpeace International, *Slaughtering the Amazon*, July 2009.
13. Robert Engelman, "Population and Climate Change—Can We Talk?" Island Press Blog, May 28, 2008.
14. Quoted in Kerri Smith, "The Population Problem," *Nature Reports Climate Change*, vol. 2, June 2008. www. nature.com.

# CHAPTER 4

# Climate Change Impacts on Population and Resources

Examples of ways the climate has influenced the human population abound throughout world history. Chapter 1 describes how the climate contributed to a transition from hunting and gathering to agrarian societies thousands of years ago. In another example, anthropologists generally believe that a cooling period twenty thousand years ago opened up a land bridge in the Bering Sea, which allowed the ancestors of American Indians to migrate from Siberia to Alaska. These examples illustrate how climate has affected the way human populations obtained food, where they lived, and how they moved from one location to another.

Scientists believe that modern-day climate change also may affect the human population in significant ways. Depending upon the magnitude of the increase in global temperature, climate change is expected to cause sea-level rise, as well as droughts, floods, and other extreme weather events in many places around the world. These events will be felt the most at the regional level, and their impact will vary based on the regional population's ability to respond and adapt. In general, people living in dry or coastal areas of developing countries will feel more climate change impacts than people living inland or in more temperate and industrialized countries.

## Water-Related Impacts

Many climate change impacts on the human population are expected to revolve around water—either too little water or too much. Climate change is expected to cause significant changes to the hydrological cycle, which describes the continuous movement of water on, above, and below the surface of the earth due to evaporation, precipitation, and infiltration. The Intergovernmental Panel on Climate Change (IPCC) asserts that climate change has already affected the hydrological cycle. According to the IPCC, observed increases in precipitation in northern latitudes and decreases in precipitation near the equator are likely to have been caused by climate change. According to climate models used to predict future potential impacts of a warmed planet, precipitation and other hydrological changes will increase in scale and intensity for each degree Celsius the global average temperature increases. These climate models suggest that many regions are expected to see decreased surface water and groundwater supplies, frequent and intense droughts and floods, and extreme weather events. All of these events will affect the regional population. According to the United Nations, "Water is the primary medium through which climate change influences the Earth's ecosystems and therefore people's livelihoods and well-being."[1]

People who live in dry locations are likely to suffer significant impacts from climate change, as already stressed water supplies become scarcer. Arid areas, such as deserts, and semiarid areas cover about one third of the earth's land area. Arid areas can be found on all continents. Even the continent of Antarctica has polar deserts, which are considered dry because all moisture is locked up in ice. There are arid areas in the Middle East; North Africa and portions of southern Africa; the southwestern United States and Mexico; the Australian outback; northern China and Mongolia; and portions of Chile, Argentina, and Peru in South America. Semiarid regions include northeastern Brazil and the Mediterranean basin.

Population growth and increased use of water are already putting mounting pressure on water resources in dry areas. Water shortages are already occurring in large parts of Australia, Asia, Africa, and in the southwestern United States. As the

## AREAS VULNERABLE TO CLIMATE-RELATED WATER CHALLENGES

**Water stress indicator:** withdrawal to availability ratio*

- 0–0.1 (no stress)
- 0.1–0.2 (low stress)
- 0.2–0.4 (mid stress)
- 0.4–0.8 (high stress)
- 0.8+ (very high stress)

*Water withdrawal: water used for irrigation, livestock, domestic and industrial purposes (2000). Water availability: average annual water availability based on the 30-year period, 1961–1990.

Source: B.C. Bates, Z.W. Kundzewicz, S. Wu, and J.P. Palutikof, eds., "Climate Change and Water," *Technical Paper of the Intergovernmental Panel on Climate Change*, Geneva: IPCC, 2008.

population grows in these areas, the need for water grows more intense. Additionally, per capita, or per person, water use is increasing. According to the United Nations' Food and Agriculture Organization, water use has been growing at more than twice the rate of population increase in the last century. Freshwater sources in dry areas include rainfall, which is minimal; seasonal surface water primarily from snow- or glacier-fed river basins;

and groundwater. Groundwater plays an important role in supplying water for agriculture in arid and semiarid regions. Over
the last several years water wells have gotten deeper and deeper,
causing many underground aquifers to become overdrawn.

In the desert southwest of the United States, scarce water
resources are under huge pressure from large populations living in Arizona, California, and Nevada. In recent years, many of

this area's water sources have been at their lowest levels. These sources include Sierra Nevada snowpack, which provides most of the water for northern California; Lake Mead, which provides nearly all the water for Las Vegas; and the Colorado River, which is fed mostly by snowmelt from the Rocky Mountains and supplies water to some thirty million people in Colorado, Utah, Wyoming, New Mexico, Arizona, Nevada, and California. Populations in many of these states have escalated in recent years. Between 2000 and 2009, the US Census Bureau estimates that Arizona and Nevada had their populations rise by about 30 percent, which was more than three times the national average. Industry, agriculture, and households are all vying for the same water resources. The increasing use of water for agriculture and computer industries caused California officials in 2009 to warn that the state is facing "the worst drought in modern history."[2]

In other dry parts of the world, particularly poor developing countries, water shortages are not only an issue of water availability, but also an issue of access and poverty. According to the organization Water.org, almost one billion people lack access to safe drinking water and 2.5 billion do not have improved sanitation. These billions of people obtain water for drinking, cooking, and other uses directly from a river or pond, an unprotected well or spring, or from unscrupulous water vendors. Moreover, they do not have toilets.

People who do not have access to improved water and sanitation are at risk for a multitude of diseases. Unimproved drinking water has a high probability of being contaminated with pathogenic bacteria and viruses. Because of a lack of sanitation facilities, human excrement, loaded with pathogenic microorganisms, contaminates the same rivers and ponds people use for drinking. Drinking-water contamination also occurs because of a lack of water for proper hygiene, such as simple hand washing—germs are transferred from a person's hands to water when it is collected and stored. Easily preventable, infectious waterborne diseases such as cholera, typhoid, guinea worm disease, and diarrhea are responsible for 80 percent of illnesses and deaths in the developing world, with children being particularly vulnerable. Cholera, for example—a dreaded disease in centuries past—is now extremely rare in the United States and other developed countries, but cholera epidemics are common in sub-Saharan Africa.

Climate change will exacerbate the water shortages felt across the world. Scientists studying climate change impacts for the US National Academy of Sciences found that precipitation amounts and stream flow, a key index of freshwater availability, will experience changes on the order of 5 percent to 15 percent per degree of warming in many areas. Arid and semiarid regions, including the southwestern United States, are expected to see the greatest water supply decreases. According to the IPCC, the populations that will be most affected by climate change with respect to water services are ones located in the already water-stressed basins of Africa, the Mediterranean region, the Near East, southern Asia, northern China, Australia, the United States, central and northern Mexico, northeastern Brazil, and the west coast of South America. Groups particularly at risk will be populations living in megacities, rural areas strongly dependent on groundwater, on small islands, and in glacier- or snowmelt-fed basins (more than one sixth of the world's population lives in snowmelt basins). Problems will be more critical in economically depressed areas, where water stress will be worsened by socioeconomic factors.

## The Significant Impact of Drought

Droughts, prolonged periods of reduced moisture, are among the most significant water-related impacts of climate change. Droughts can be defined in several ways, depending on their effects and whether they are caused by too little rain or dwindling supplies of surface water or groundwater. The National Oceanic and Atmospheric Administration's National Weather Service explains the following about drought: "While much of the weather that we experience is brief and short-lived, drought is a more gradual phenomenon, slowly taking hold of an area and tightening its grip with time. In severe cases, drought can last for many years, and [it] can have devastating effects on agriculture and water supplies."[3] These effects reverberate throughout society: farmers lose income if a drought destroys crops or makes it harder to obtain water or feed for animals. Reduced supplies of grains, fruits, vegetables, and meat result in higher and higher food prices for consumers. People already struggling because of poverty find it increasingly difficult to obtain adequate food supplies. Many people, particularly the young, are vulnerable to the effects of reduced caloric intake, nutritional deficiencies, and malnutrition. Drought can also lead to disease epidemics through contaminated drinking water, inadequate sanitation, or poor hygiene.

When droughts are severe, thirst, hunger, and famine can result. A famine is a widespread scarcity of food that causes malnutrition, starvation, increased susceptibility to disease, and increased mortality. Famines are common events throughout history, occurring regularly in parts of Africa, China, and Russia. Droughts and other weather-related events contribute to or are generally the main cause of famine. The Great Chinese Famine (1958–1961), during which some thirty million people starved to death, was caused primarily by drought.[4] Food distribution problems, tyrannical governments, and warfare also contribute to or cause famines, however. For instance, many historians believe that Soviet leader Joseph Stalin intentionally caused a famine that killed millions of Ukrainians between 1932 and 1933, in order to punish an emerging Ukrainian nationalism.

Droughts can contribute to urbanization, as people flee the countryside in search of water, food, and work. When droughts last for extended periods of time, rural residents, who have depended on growing their own food, may find no alternative but to move and try to make a living elsewhere. A drought in northeastern Syria in 2010 caused some three hundred thousand families to flee the countryside and move to Damascus and other cities. The United Nations called it the "largest internal displacement in the Middle East in recent years."[5] A wheat farmer named Ahmed Abu Hamed Mohieddin, who was widely quoted in newspaper accounts of the drought, said, "Our wells are dry and the rains don't come. We cannot depend on God's will for our crops. We come to the city, where the money is."[6]

Droughts affect not only people but can also cause permanent impacts to land. Persistent drought, along with human exploitation—such as through livestock grazing, intense agriculture, and deforestation—can degrade semiarid lands into hot dry deserts in a process known as desertification. Semiarid lands sustain a fragile ecosystem of plants and animals that have adapted to the climate. When these lands are continuously pushed to their limits, any upset in the balance of water availability can permanently transform them into arid lands. Desertification endangers wildlife, reduces biodiversity, and increases difficulties for people trying to make a living on the land. According to the IPCC, many semiarid regions in southern Africa, parts of Mexico, and Brazil are at risk of desertification.[7]

## Floods: Swift and Deadly

Whereas droughts are prolonged climatic events, or creeping disasters, floods—another major water-related impact of climate change—are generally swifter and can be deadlier. Floods happen when the water draining from a watershed (the area of land that water flows across or under as it makes its way to a stream, river, or lake), whether from rainfall or melting snow or ice, exceeds the capacity of the river or stream channel to hold it. In hilly and mountainous areas, floods can be rapid, deep, and dangerous. In relatively flat areas, land may stay covered with shallow, slow-moving water for days or even weeks. Climate change is expected to increase the risk of flooding from heavy rainfall and the melting of glaciers and mountain snows.

Floods are among the most costly natural disasters. Floods can kill people and animals; wash away villages; destroy crops, water sources, and property; and cause people to migrate to escape the rising waters. As with drought, floods can lead to contaminated water and lack of sanitation, which increases the risk of communicable diseases, particularly in developing countries, where drinking water supplies and sanitation are inadequate to begin with and public health infrastructure is unlikely to exist. People seeking respite from the floods who end up living in overcrowded shelters are also at a higher risk of contracting communicable diseases.

Floods have occurred throughout history and are common in many areas of the world, particularly in Asia. China has been particularly hard hit by floods. The worst flood in human history occurred in 1887, when the Yellow River overran the dikes in Henan province. That flood covered 50,000 square miles. It inundated eleven large towns and hundreds of villages. Nine hundred thousand people died, and two million were left homeless. The river has overflowed at other times as well, causing many Westerners to dub it "China's Sorrow."

According to the IPCC, Asia has already begun experiencing more frequent floods and landslides due to climate change. The events of the summer of 2010 seem to support this observation. Heavy rains caused the worst flooding in Pakistan's history. Floods that began in the mountainous areas of northern Pakistan after a two-day period of heavy rainfall in late July triggered a torrent of water that flowed down the Indus River. The inundated river became swollen with water, which overran its banks, affecting more and more people as the river flowed south toward the Arabian Sea. According to the United Nations, millions of people were affected, and more than a thousand were left dead. Floodwaters submerged villages and cities, destroyed crops, and killed livestock. While Pakistan was experiencing flooding, people in northern China were experiencing that region's most severe mudslides in decades. According to the Chinese news agency Xinhua, more than one thousand people perished in China's Gansu province as torrential rains caused massive mudslides in August 2010.

## Storms and Sea-Level Rise

Other expected consequences of climate change include an increase in the frequency and severity of tropical storms as well as rising sea levels. Both such events can cause floods and significantly impact people living in low-lying and coastal regions. As global temperatures rise, the ocean gets warmer and expands, causing sea levels to rise. In addition, the melting of mountain glaciers and land-based ice sheets is expected to raise sea levels. Rising sea levels make coastal areas more vulnerable to storms and particularly to flooding. The IPCC predicts that warming tropical seas will increase the frequency and strength of hurricanes and other tropical storms, which feed off the warm water. Storms can bring torrential rains that wreak havoc on coastal areas and cause flooding. Densely populated, low-lying areas, such as large river deltas and small islands, are at the greatest risk from flooding. Many of these areas are found in Asia, such as the Ganges River Delta, the Mekong River Delta, and islands in the South Pacific.

The Organization for Economic Cooperation and Development (OECD), an international organization with thirty member countries, including the United States, studied the current and future exposure of the world's large port cities (ones with more than a million residents) to coastal flooding due to storm surge and damage from high winds. According to the OECD, the ten cities with the highest population exposure to storms and floods in 2005 were Mumbai, India; Guangzhou, China; Shanghai, China; Miami; Ho Chi Minh City (formerly Saigon), Vietnam; Kolkata (Calcutta), India; New York; Osaka-Kobe, Japan; Alexandria, Egypt; and New Orleans. These cities are almost equally split between developed and developing countries and include an estimated 40 million people. The OECD predicts that by 2070, the combined effects of climate change, subsidence (a circumstance in which the ground level gradually falls due to natural or human causes, such as the extraction of groundwater), population growth, and urbanization will increase the number of exposed people to approximately 150 million and shift the exposed cities unequally to the developing world. In the future, residents of Kolkata and Mumbai (formerly Bombay) are expected to become the two most exposed populations, followed by Dhaka, which is the largest city in Bangladesh, then Guangzhou; Ho Chi Minh City; Shanghai; Bangkok, Thailand; and Rangoon, Burma—all of which are in Asia. Miami is ninth on the list, followed by Hai Phong, a large city in Vietnam. Most of the vulnerable large-city populations are found in Asia and in deltaic settings, which have higher coastal flood risk because they tend to be at lower elevations.[8]

Many people in Miami, Florida, the only US city on the 2070 most vulnerable coastal cities list, are worried about climate

*Hurricane Wilma flooded Miami, Florida, in 2005. Rises in sea level due to climate change may result in the loss of 10 percent of Florida's land area.*  Roberto Schmidt/AFP/Getty Images.

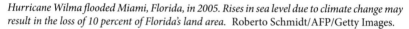

change impacts. The biggest concern for Miami is the threat of sea-level rise. According to the US Environmental Protection Agency, sea level in southern Florida is likely to rise 20 inches (50 cm) above 1990 levels by 2100.[9] As sea level rises, coastal areas in Florida could be inundated.

Frank Ackerman, a senior economist at the Stockholm Environment Institute who has studied sea-level rise in Florida, was interviewed for a story about climate change by National Public Radio (NPR) in December 2009. Ackerman believes that Florida will see its sea-level rise more than 20 inches (50 cm) before the end of the century. He thinks it will soar by 27 inches (69 cm) by 2070 and cause the loss of 10 percent of Florida's land area and the homes of 1.5 million people. "It's like our map of the area vulnerable to 27 inches of sea level rise looks like someone took a razor to the state right above Miami and sliced off everything below that," Ackerman told NPR. Ackerman has also mapped the impact a 5-foot sea-level rise would have on South Florida. On that map, Miami Beach and nearly all of the Keys are gone, and most of Miami-Dade County and much of Broward County are underwater. According to Ackerman, "It's going to just become increasingly risky to live in a place like Florida."[10] Florida officials are among groups who are concerned about climate change. Miami-Dade County created a task force to make recommendations on steps the county can take to mitigate climate change and prepare for its impacts, particularly sea-level rise. Among the task force's recommendations are to plan new development in high-ground areas, create local infrastructure outside coastal zone areas, and acquire undeveloped lands that can serve as buffer zones and keep them from being developed in the future.

While the planning and preparation of Florida officials may help people in Miami and the surrounding region to adapt to sea-level rise, people living in many of the other coastal cities on the 2070 vulnerability list might not be so lucky. It is not clear whether such cities as Kolkata, Dhaka, Bangkok, or Rangoon are planning for sea-level rise. Without a plan, people in these cities may face an uncertain future.

## Food Sources in a Warming Climate

As described above, water's impact on food sources in a warming climate can be devastating. Both too little and too much water threaten crops and livestock. Apart from the important role water plays, food sources, particularly agricultural crops, are also directly affected by climate change. The direct impact of climate change on agriculture is highly variable and is largely dependent on regional conditions. In some areas climate change may even have a favorable impact on agriculture. For instance, an increase in average temperature can lengthen the growing season in some

## The "Sting" of Climate Change

One area where the negative impacts of climate change are already being felt, and felt most sharply in the Indonesian archipelago and the small island states of the South Pacific, is in the spread of the deadly mosquito-borne diseases . . . malaria and dengue. Each year there are up to half a billion new cases of malaria and as many as two million deaths globally, mostly children. Dengue is also a significant health problem, with an estimated 50–100 million cases of dengue fever annually and approximately 25,000 deaths. Experts agree that this number is rising every year.

Unfortunately, mosquitoes are very sensitive to changes in climate. Warmer conditions allow the mosquitoes and the malaria parasite itself to develop and grow more quickly, while wetter conditions let mosquitoes live longer and breed more prolifically.

Projections of the impact of climate change on malaria and dengue are truly eye-opening. Early modeling studies predict that malaria prevalence may be 1.8 to 4.8 times greater in 2050 than in 1990. The share of the world's people living in malaria-endemic zones may grow from 45% to 60% by the end of the century, with 'fringe zones' like northern Aus-

temperate regions. Additionally, increased levels of carbon dioxide—the main cause of climate change—can spur plant growth. Carbon dioxide is a key ingredient in photosynthesis, the process through which plants produce carbohydrates and grow. The idea that increased atmospheric carbon dioxide can have beneficial effects is known as the fertilization effect, because higher carbon dioxide concentrations in the atmosphere can literally fertilize plant growth. The fertilization effect is not expected to occur in tropical or developing world regions, but it may occur in countries located in temperate areas, like the United States.

Agriculture in the United States and other industrialized countries will be less vulnerable to climate change than agriculture in developing nations. IPCC climate models show that

tralia at the forefront of this expansion. By 2085, it is estimated that 52% of the world's population, about 5.2 billion people, will be living in areas at risk of dengue.

Beyond these alarming headline numbers, one of the greatest problems in addressing the intersection of climate change and mosquito-borne diseases is the significant uncertainty over how much and how quickly climates will change and how much of the observed and predicted increases in malaria and dengue can be directly linked back to climate change. However, many experts agree that early warning signs already exist. Recent data suggest that since the 1970s climate change has contributed to 150,000 more deaths every year from disease, with over half of these deaths occurring in Asia. In Africa, as well as Papua New Guinea, increases in average temperature have contributed to the spread of malaria into previously malaria-free zones, particularly in highland areas. It has been estimated that in the African highlands a mere half-degree rise in temperature translates into a 30–100% increase in the number of mosquitoes able to transmit malaria.

SOURCE: Sarah Potter, "The Sting of Climate Change," *Lowy Institute Policy Brief*, November 2008. www.lowyinstitute.org/Publication.asp?pid=926.

moderate regional temperature increases of 1°C to 3°C (about 1.8°F to 5.4°F) will significantly reduce crop yields in tropical areas and most developing countries, while temperate and largely industrial countries may see small positive effects. For temperature increases above 3°C, yield losses are expected to occur everywhere and be particularly severe in tropical regions.[11] Climate change is expected to cause yield declines for the most important crops in developing countries. South Asia and sub-Saharan Africa will be among the hardest hit.

Climate change threatens to increase the number of undernourished and malnourished people in the developing world, particularly in Asia and sub-Saharan Africa. Undernourishment and malnourishment are major problems in the developing world even without climate change impacts. According to 2009 statistics from the United Nations Food and Agriculture Organization, there are more than a billion hungry people in the world and 915 million of them are in developing countries. The majority of these hungry people live in rural areas, mainly in the villages of South Asia and sub-Saharan Africa. They are overwhelmingly dependent on agriculture for their food and have no alternative source of income or employment, circumstances that make them highly vulnerable to climate change.

South Asia and southern Africa have been identified as "hunger hotspots" in the context of climate change.[12] Crop yields of corn, which is the most important crop in southern Africa, are expected to be 30 percent lower in 2030 than they were in 1990. In South Asia, many crops have more than a 75 percent chance of incurring losses from climate change. They include wheat, rapeseed, rice, millet, and maize. Considering the increased food needs population growth will cause for the developing world, the likelihood of these declines is alarming. The International Food Policy Research Institute estimates that climate-change-related crop losses, along with rising food prices, will cause the number of malnourished children younger than five to increase by twenty-five million by 2050.

## Climate-Induced Migration

Human migration may be one of the most important impacts of climate change. Throughout history, large numbers of humans have left their homes and relocated to other areas or countries, both voluntarily and involuntarily. The migration of large numbers of Europeans and Asians to the United States in the late nineteenth and early twentieth centuries is an example of voluntary migration. The large numbers of Sudanese people fleeing ethnic violence in Darfur, Sudan, early in the twenty-first century is an example of involuntary, or forced, migration.

*Climate change has the potential to produce hundreds of millions of environmental refugees through droughts, floods, and extreme weather events.*

According to the United Nations, climate change might cause both forced and voluntary migration. The United Nations expects reductions in the viability of agriculture and water availability, increases in extreme weather events, and sea-level rise to generate massive displacements. Some people will move voluntarily before crises exist. But many poor have no alternative except to wait until disaster strikes and then be forced to flee.

Climate change has the potential to produce hundreds of millions of environmental refugees through droughts, floods, and extreme weather events. An environmental refugee is a person displaced from his or her home because of environmental causes—notably land loss and degradation—and natural disaster. Like the wheat farmer from Syria who took his family to the city to escape drought, many people in the developing world flee natural disasters or land degradation. The population of Dhaka, Bangladesh, identified by the OECD as one of the coastal cities most vulnerable to sea-level rise and storms, is skyrocketing as the city receives environmental refugees fleeing natural disasters such as cyclones, landslides, and floods. Dhaka is the fastest-growing megacity in the world, according to the World Bank. At least twelve million people live in Dhaka, and it receives more than four hundred thousand newcomers each year.[13]

Many scientists and humanitarian organizations have tried to estimate the number of environmental refugees that climate change will create. The most widely cited figure is that given by Oxford University professor Norman Myers. He predicts that as many as two hundred million people could be displaced by disruptions of monsoon systems and other rainfall regimes, by droughts of unprecedented severity and duration, and by sea-level rise and coastal flooding—all caused by climate change. This number would represent almost a tenfold increase over the number of environmental refugees documented in 1995.[14] Scientists are trying to provide more data on the potential for climate-induced migration in order to help policy makers draft appropriate governmental policies and to help aid groups plan for what may be a massive number of environmental refugees.

## The Vulnerability Gulf

The future existence of a vulnerability gulf, or climate change divide, becomes apparent when analyzing the impacts of climate change on human populations. The most severe and frequent impacts of climate change—the most intense and frequent droughts, floods, and storms, and the highest sea-level rise—will occur in developing countries. At the same time, expanding populations, scarce resources, little or no infrastructure, poverty, and unstable governments mean developing countries will be the least able to adapt to such events.

Industrialized countries, generally due to their temperate climates, are expected to see fewer impacts than the developing world, however, and may even see some benefits from climate change. Industrialized countries will also be better able to adapt to climate change. The vulnerability gulf illustrates the disparate effects of climate change and highlights the overarching divide between rich and poor, developed and developing, industrial and nonindustrial countries.

## Notes

1. UN-Water, "Fact Sheet: 'Climate Change Adaptation Is Mainly About Water,'" www. unwater.org.
2. Claire Baldwin, "Snow Study Shows California Faces Historic Drought," Reuters, January 29, 2009.
3. National Oceanic and Atmospheric Administration, "Drought: The Creeping Disaster," March 15, 2001.
4. Vaclav Smil, "China's Great Famine: 40 Years Later," *British Medical Journal*, December 18, 1999.
5. SOS Children's Villages, "Thousands of Families Flee Syria Drought," March 3, 2010. www.soschildrensvillages.org.uk.
6. Quoted in Daniel Williams, "Drought Threatens Syria Economy as Refugees Flee Parched Farms," *BusinessWeek*, March 1, 2010.
7. B.C. Bates et al., eds., *2008: Climate Change and Water. Technical Paper of the IPCC*, Geneva, June 2008.
8. Organisation for Economic Co-operation and Development, "Ranking of the World's Cities Most Exposed to Coastal Flooding Today and in the Future," 2007.
9. Environmental Protection Agency, *Climate Change, Wildlife, and Wildlands: Case Study—Everglades and South Florida*.
10. Quoted in Greg Allen, "Florida Faces Drastic Change From Sea Level Rise," NPR, December 11, 2009.
11. W.E. Easterling et al., "Food, Fibre and Forest Products," *Climate Change 2007: Impacts, Adaptation and Vulnerability. Contribution of Working Group II to the Fourth Assessment Report of the IPCC*, M.L. Parry et al., eds., New York: Cambridge University Press, 2007.
12. David Lobell et al., "Prioritizing Climate Change Adaptation Needs for Food Security to 2030," *Stanford University Program on Food Security and the Environment—Policy Brief*, 2008.
13. Joanna Kakissis, "Environmental Refugees Unable to Return Home," *New York Times*, January 3, 2010.
14. Norman Myers, "Environmental Refugees: An Emergent Security Issue," Organization for Economic Security and Cooperation in Europe, 13th Economic Forum: Prague, May 2005.

# CHAPTER 5

# Vulnerable Populations

People everywhere will feel the impacts of climate change. Yet some will be affected more than others. People living in certain regions of the world and people who have limited means to adapt to climate change are the most vulnerable.

## Defining Climate Change Vulnerability

People, wildlife, plants, or entire ecosystems can be vulnerable to climate change. The Intergovernmental Panel on Climate Change (IPCC) defines climate change vulnerability as "the degree to which a system is susceptible to, and unable to cope with, adverse effects of climate change, including climate variability and extremes. Vulnerability is a function of the character, magnitude, and rate of climate change and variation to which a system is exposed, its sensitivity, and its adaptive capacity."[1]

By virtue of their location, current climate, elevation, proximity to water, and other factors, certain regions of the world are more susceptible to climate change's adverse effects than others. In general, people living in dry areas and coastal areas are most susceptible. As discussed in Chapter 4, people in these areas are most at risk from droughts, floods, sea-level rise, storms, and agricultural effects.

Vulnerability also refers to the ability of systems to cope with or adapt to climate change impacts. In general, the capability of human populations to adapt to climate change is based on the social, economic, governmental, or infrastructure-related resources available to them. These resources could be things that help the population become resilient to climate change impacts before they happen, such as building codes that require sturdier buildings or improved roads that allow aid to reach a disaster area. These resources could also be things that help people after a climate change event occurs, such as financial resources that allow people to move to a new location, educational resources that provide people with the opportunity to learn new skills if their source of income or way of life is destroyed, or government resources that provide emergency assistance such as health care, food, water, and shelter.

*People who already struggle with poverty, disease, and hunger may be pushed to the brink by climate change impacts, such as the destruction of their only source of food or water.*

A population's vulnerability to climate change is affected by its existing standard of living and quality of life. People who already struggle with poverty, disease, and hunger may be pushed to the brink by climate change impacts, such as the destruction of their only source of food or water. Poor people generally have far fewer of the resources described above and, as a result, are most vulnerable to climate change. People who are already suffering from diseases, such as cholera or diarrheal infections, because of a lack of access to safe water and improved sanitation may be in a weakened state and have a harder time responding to climate change. People who live in countries with an unstable or tyrannical government or in areas with conflicts such as war or terrorism, often live in an atmosphere of fear. Under climate change impacts, these people may become easy prey for militants and insurgents.

The fate of millions of flood victims in Pakistan in the summer of 2010 provides an example of a population vulnerable to climate change impacts. In late July and August 2010, heavy rainfall triggered extensive flooding of the mighty Indus River, which cuts a large swath down the middle of Pakistan. The floods killed more than a thousand people, destroyed millions of homes, and devastated thousands of acres of crop land. Government efforts to get food, water, and aid supplies to the people who needed them were largely ineffective during the early days of the disaster because the country's transportation infrastructure was inadequate. Millions of people who fled their homes were stranded along the roadside, without shelter, in sweltering temperatures for days. Individuals who made it to overcrowded shelters were

## Climate Change Threatens to Leave Many People "Stateless"

Sinking island states present one of the most dramatic scenarios of the impact of climate change. The entire populations of low-lying states such as the Maldives, Tuvalu, Kiribati and the Marshall Islands may in the future be obliged to leave their own country as a result of climate change. Moreover, the existence of their state as such may be threatened. Entire populations of affected states could thus become stateless.

Article 1 of the *1954 Convention Relating to the Status of Stateless Persons* defines a stateless person as a person who is not considered a national by any state under the operation of its law. Should a state cease to exist, citizenship of that state would cease, as there would no longer be a state of which a person could be a national. The question is then the extent to which climate change could affect statehood.

Should the entire territory of a state be permanently submerged, inevitably there could be no permanent population attached to it or a government in control of it. The loss of all territory has been cited most frequently as [possible grounds] for loss of statehood. It appears, how-at risk of developing cholera, diarrheal infections, or other waterborne or respiratory diseases. According to a story in the *New York Times*, of the more than 15 million people the floods are estimated to have affected, only about 1.2 million had access to safe water supplies. In the most inundated areas, 200 of 1,167 health facilities—including several hospitals—were damaged. Reports of respiratory tract infections and fatal diarrhea were common.[2]

In a story reported by the news agency Reuters, the floods caused a Pakistani woman, named Said, to flee her mud-brick home along with her seven children. At the time of the report, Said and her children had been at an overcrowded shelter for eighteen days. Her one-year-old child, Naeema, had been born

ever, unlikely to occur before the end of the century, even with the upwardly revised rates in rising sea-levels announced by scientists recently.

A threat to statehood may nonetheless arise far earlier. It is projected that the number and severity of extreme events such as storms and flooding will increase considerably. Extensive loss of fresh water and arable land due to contamination and seepage is expected. As well, destruction of the economic base is additionally expected due to erosion, as well as damage to corals and fishing grounds due to rising sea levels and global warming. The Intergovernmental Panel on Climate Change has thus indicated that rapid sea-level rise that inundates islands and coastal settlements is likely to limit adaptation possibilities, with potential options being limited to migration. It has also confirmed that rising sea-levels are unavoidable.

Low-lying island states are thus very likely to be entirely uninhabitable long before their full submersion, causing entire populations and the governments to be externally displaced. Unless territory could be protected or territory was ceded by another state, the exile of the population and the government would presumably be permanent.

SOURCE: United Nations High Commissioner for Refugees, *Submission: Climate Change and Statelessness: An Overview*, 6th Session of the Ad Hoc Working Group on Long-Term Cooperative Action, Bonn, Germany, June 1–12, 2009.

with a hole in her heart and was in a weakened state. But there was no doctor available. According to the Reuters story:

> Naeema sleeps on a mat in sweltering heat at a fly-infested camp, with no fan, no chance of seeing a cardiologist anytime soon and at risk of catching other potentially fatal diseases in cramped, un-hygienic conditions. "Who will treat her? The doctors said she has a hole in the wall of her heart," said Said, also worried because one of her six other children has fallen ill after the floods.[3]

A comparison between two earthquakes, both of which occurred in 2010, illustrates the impact a population's vulnerability can have on the number of people who are harmed by a disaster. On January 12, 2010, a 7.0 magnitude earthquake struck Haiti near its capital, Port-au-Prince. According to Haitian government reports, an estimated 230,000 people died, 300,000 were injured, and a million lost their homes. About a month and a half later, on February 27, a much stronger 8.8 magnitude earthquake occurred off the coast of the Maule region of Chile. An estimated 521 people were killed. The Chilean earthquake was five hundred times stronger than the one that ravaged Haiti. Seismologists estimated that the earthquake was so powerful that it may have shortened the length of the planetary day by 1.26 microseconds. Generally, however, the people in Chile fared better than those in Haiti, due in part to a better government emergency response and stricter building codes that resulted in sturdier structures. On the blog Ecopolity.com, Brazilian scientist and researcher Sergio Abranches noted the importance of a population's vulnerability in determining the human toll of a disaster, such as those that may occur because of climate change:

> There are no natural catastrophes. Catastrophes are always human-made. They are social, not physical, phenomena. A catastrophe happens when a high-intensity or extreme natural event meets a vulnerable population, with a weak, unprepared or reckless government.[4]

## Marginalized Groups Are Especially Vulnerable

Vulnerability varies within countries, within communities, and even within households. Marginalized groups, such as the poor, women, children, the disabled, and the elderly are particularly at risk of experiencing adverse outcomes because of climate change. These groups are vulnerable because they generally lack control in the decisions that affect their lives. These marginalized groups are the most vulnerable of the vulnerable populations.

The potential vulnerability of marginalized groups to climate change can be illustrated by the experience of the poor, the sick, and the elderly during Hurricane Katrina in 2005. The day before the hurricane struck the coast of Louisiana, the city of New Orleans had ordered a mandatory evacuation. No provisions were made to evacuate sick, disabled, or elderly residents, or the poor who relied on public transportation for mobility, however. Many from these marginalized groups ended up being stranded at the New Orleans Superdome without adequate food, water, toilet facilities, or police to keep order. One of the most tragic incidents of Hurricane Katrina occurred when thirty-five nursing home patients drowned. The elderly patients at St. Rita's Nursing Home did not have the power to leave the city on their own, so when St. Rita's operators decided not to evacuate, the elderly patients had no choice but to stay. When the storm waters inundated the nursing home, dozens of residents—many of whom were bedridden—were trapped and drowned.

## Poverty and Inequality Contribute to the Vulnerability of Women

According to the United Nations Population Fund (UNFPA) and the Women's Environment & Development Organization (WEDO), women are particularly vulnerable to climate change because the majority of the world's poor are women and they face inequality in several aspects of their lives. In many parts of the world, women cannot obtain financial credit or own land. Girls

*Women were much more likely than men to be victims of the tsunami that struck South Asia in 2004.* AP/Saurabh Das.

are also far less likely to attend school than boys because girls are responsible for important household chores, such as obtaining drinking water and energy fuels, such as wood, agricultural residues, or dung. These responsibilities stay with women throughout their lives. In 2008 the fifty-second meeting of the United Nations Convention on the Status of Women identified climate change impacts and the particular vulnerability of women as an emerging issue requiring attention. The Moderator's Summary of the meeting provided an example of the impact climate change can have on women in Africa:

> In Africa, women have primary responsibility for food security, household water supply, and the provision of energy for cooking and heating. Conditions such as drought, deforestation and erratic rainfall have a disproportionate negative affect on their ability to carry out these duties. As climate change

causes African women to work harder to secure these basic resources, they have less time to secure an education or earn an income. Girls are more likely than boys to drop out of school to help their mothers gather fuel, wood and water.[5]

According to UNFPA and WEDO, death tolls from natural disasters reflect women's heightened vulnerability. Women and children are fourteen times more likely than men to die during natural disasters. More than 70 percent of the dead from the 2004 Asian tsunami were women, and Hurricane Katrina, which struck New Orleans in 2005, predominantly affected African American women. UNFPA and WEDO say that poor and disadvantaged women, whether living in developing or developed countries, are disproportionately vulnerable to environmental disasters.

## Hot Spots

In 2008 the humanitarian group CARE undertook a mapping study to identify climate change "hot spots," countries and regions of the world most vulnerable to climate change impacts. The organization looked specifically at regions most susceptible to floods, cyclones, and droughts. Flood-risk hot spots were identified in Africa (particularly in the Sahel, the Horn of Africa, the Great Lakes region, central Africa and southeastern Africa); central, South and southeastern Asia; and Central America and the western part of South America. Drought-risk hot spots are mainly located in sub-Saharan Africa, South Asia (particularly Afghanistan, Pakistan, and parts of India) and Southeast Asia (notably Burma, Vietnam, and Indonesia). Cyclone-risk hot spots include Mozambique, Madagascar, Central America, Bangladesh, parts of India, Vietnam, and several other southeastern Asian countries. According to CARE researchers, southeastern Africa and parts of southern and southeastern Asia are vulnerable to all three climate change hazards: floods, droughts, and cyclones.

**Cyclone hot spot**

**Flood hot spot**

**Drought hot spot**

This map shows risk hot spots for three major climate-related hazards—flood, cyclones, and drought. Risk hotspots are defined as areas where high human vulnerability coincides with the distribution of weather-related hazards.

Source: Andrew Thow and Mark de Blois, "Climate Change and Human Vulnerability: Mapping Emerging Trends and Risk Hotspots for Humanitarian Actors, Summary for Decision Makers," *Maplecroft Report to the UN Office for Coordination of Humanitarian Affairs with CARE*, March 2008. www.careclimatechange.org/files/reports/Human_Implications_DiscussionPaper.pdf.

*Bangladesh:* Bangladesh, identified as a hot spot in the CARE study, is one of the countries most often described as being vulnerable to climate change. Bangladesh is extremely susceptible to flooding. Most of the country sits only a few meters above sea level and lies in the flood plains of three major Himalayan rivers, the Ganges, Meghna, and the Brahmaputra. The rivers have a tendency to overrun their banks due to monsoons and large volumes of melting snow from the Himalayan mountains, which

feed the rivers. According to climate scientists, more than 20 percent of Bangladesh is flooded every year, and during extreme events up to 70 percent of the country is inundated with water.

Flooding is a fact of life for Bangladeshis and one that threatens to become even more common under climate change. Bangladesh is expected to undergo more frequent cyclones and more rainfall, both of which have potential to cause major flooding, loss of life, and damage far inland. The low-lying nation is also expected to be affected by sea-level rise, which can increase the extent and magnitude of flooding and destroy freshwater sources and agricultural crops. Saltwater intrusion caused by rising seas reaching farther and farther inland is already destroying crops in Bangladesh. In a story published in January 2009 in *Nature Reports Climate Change*, environmental journalist Mason Inman interviewed Bangladeshi villagers Santosh Kumar Gain and Matthew Digbijoy Nath about the saltwater intrusion. According to Gain, "Sometimes when we plant crops, after 10 or 15 days, if there is no growth we pull up the plants and see there is no growth of the roots. We are really tired of this." Nath explains, "This area was all [rice] paddy before. Now, no paddy. The trees look nice, but the coconut trees—there are no coconuts on them. If it gets more salty here, this population will not be able to live here. No paddy, no fish. How will people live?"[6]

Bangladesh is vulnerable not only because it is expected to experience severe flooding and other events due to climate change, but also because it has a poor standard of living and is one of the countries least able to adapt. Statistics compiled by the World Bank from 2002 through 2008 reveal the vulnerability of the country. Nearly 75 percent of its people live in rural areas, and 40 percent live in poverty. Less than 50 percent of its people attend secondary school, only 7 percent go on to college, and 47 percent are illiterate. More than 40 percent of children younger than age five suffer from malnutrition.

Exacerbating Bangladesh's vulnerability is the fact that the country is very densely populated. The encroaching seas, saltwater intrusion, cyclones, flooding, and droughts caused by climate change threaten to destroy the homes of millions of people in Bangladesh. The opportunities for migration in such a densely populated country—according to the World Bank, 160 million people squeeze into an area the size of Iowa—are limited. According to journalist Inman, "there's little doubt that the future could see many climate change refugees—perhaps tens of millions—fleeing parts of Bangladesh."[7] According to

Atiq Rahman, executive director of the Bangladesh Center for Advanced Studies, "Already, climate change is having enough of an impact here that it's partly responsible for pushing some people off their land . . . I believe there are climate change refugees already."[8]

*Sub-Saharan Africa:* Sub-Saharan Africa is another hot spot identified in the CARE study. Climate change is expected to reduce the availability of water in sub-Saharan Africa, a region already experiencing water scarcity. According to a 2007 United Nations Framework Convention on Climate Change report, one third of African people already live in drought-prone areas, and 220 million are exposed to drought each year.

A dependence on agriculture, the scarcity of water, the lack of resources, and the presence of conflicts make the people of sub-Saharan Africa extremely vulnerable to climate change. Much of the population of this region, particularly in the Horn of Africa, are subsistence farmers (people who grow only enough food to feed their families) and pastoralists (nomadic people who depend on livestock). Both these groups of people are almost entirely dependent on the land for their livelihoods. Many of them lack access to public water systems or improved sanitation. They are at risk for such diseases as cholera, malaria, or HIV/AIDS. They lack access to social services, education, and health care. If the climate—through drought, flood, or storm—destroys their crops, dries up their water sources, kills their livestock and/ or destroys their homes, they generally have nowhere to turn. Compounding the situation in much of Africa and making the population even more vulnerable is the presence of war and conflict. Virgil Hawkins—author, former humanitarian worker in Africa and Asia, and an assistant professor at the Global Collaboration Center at Osaka University in Japan—calls many of the conflicts in Africa, including ones in Somalia, Ethiopia, Eritrea, and Sudan, "stealth conflicts" because they occur largely under the radar of mainstream media are and not well known to many people around the world. Nonetheless these conflicts create hardship for various populations that is only expected to worsen under climate change. In a press release issued September 21, 2009, the United Nations Food and Agriculture Organization's Food Security and Nutrition Analysis Unit warned that Somalia is facing the worst humanitarian crisis in eighteen years, with half the population—3.6 million people—in need of lifesaving assistance. This number encompasses 1.4 million people in rural areas affected by severe drought, more than 650,000 urban poor facing high food prices, and 1.3 million people who have been uprooted by violence.

In an October 2009 article in the *Los Angeles Times*, Edmund Sanders recounts the plight of herders and pastoral people in sub-Saharan Africa and indicates that many of those people, like those in Bangladesh, are already leaving their homes because of climate change. Sanders tells the story of Adam Abdi Ibrahim, whose ancestors herded cattle and goats across an unforgiving landscape in southern Somalia for years. Eventually Ibrahim had no choice but to seek refuge for himself and his family in a crowded Kenyan refugee camp. According to Sanders, "Ibrahim became the first in his clan to throw in the towel, abandoning his land and walking for a week to bring his family to this overcrowded refugee camp in Kenya. He's not fleeing warlords, Islamist insurgents or Somalia's eighteen-year civil war. He's fleeing the weather." Sanders reports that while Ibrahim is going to a refugee camp, others displaced by the climate are flooding into Kenya's larger cities, such as Nairobi. The city of Nairobi is full of refugees—people, like Ibrahim, fleeing the weather, or others fleeing violence. According to Amnesty International, more than two million people live in the slums of Nairobi. As Sanders explains, the fate of the refugees "is a reminder that behind the science, statistics and debate over global warming, climate change is already having a deep impact on Africa's poverty, security and culture."[9] This observation is true for Africa, for Bangladesh, and for many other vulnerable populations throughout the world.

# Notes

1. Intergovernmental Panel on Climate Change, *Glossary of Terms Used in the IPCC Fourth Assessment Report*, 2007.
2. Neil MacFarQuhar, "Aid for Pakistan Lags, U.N. Warns," *New York Times*, August 18, 2010.
3. Michael Georgy, "Disease Hovers over Pakistan's Flood-Stricken Children," Reuters, August 18, 2010.
4. Sergio Abranches, "There Are No Natural Disasters, Only Social Catastrophes," Ecopolity.com, March 1, 2010.
5. Moderator's Summary, "Gender Perspectives on Climate Change," *52nd session of the United Nations Commission on the Status of Women*, February 25–March 7, 2008.
6. Quoted in Mason Inman, "Where Warming Hits Hard," *Nature Reports Climate Change*, January 15, 2009.
7. Inman, "Where Warming Hits Hard."
8. Quoted in Inman, "Where Warming Hits Hard."
9. Edmund Sanders, "Fleeing Drought in the Horn of Africa," *Los Angeles Times*, October 25, 2009.

CHAPTER 6

# Climate Change and Conflicts

M any people believe that climate change has the potential to cause conflicts worldwide as people fight over dwindling resources and flee from disasters, and as governments fail. Individuals from a diverse collection of groups—humanitarian, military, environmental, and academic—are concerned that water scarcity, drought, floods, extreme weather, damaged crops, hunger, land loss, migration, and other climate change impacts will heighten tensions and even cause armed conflicts between different groups of people struggling to adapt to climate change.

## Fighting over Water

Water is a central factor in the link between climate change and conflict. Water is vital for human needs, agriculture, and industry and, as described in Chapter 4 and Chapter 5, it plays an important role in human vulnerability to climate change. According to the organization Water.org, nearly a billion people lack access to safe water and 2.5 billion lack improved sanitation. These numbers are predicted to rise as climate change affects temperatures, rainfall, and the levels of lakes, rivers, and streams.

Water has been called "blue gold," or the "oil of the twenty-first century" by various authors and scientists. In his book *Tapped Out: The Coming World Crisis in Water and What We Can Do About It*, former US congressman Paul Simon writes, "Nations fight over oil, but valuable as it is, there are substitutes for oil. There is no substitute for water. We die quickly without water, and no nation's leaders would hesitate to battle for adequate water supplies."[1]

Many scientists believe that water has already caused regional conflicts. Daniel Hillel from the University of Massachusetts has examined the role water played in contributing to conflict in the Middle East. In his 1994 book *Rivers of Eden*, Hillel traced the history of the region's conflicts—ones involving Israel, Syria, Jordan, Saudi Arabia, Palestine, Iraq, and Turkey—in the context of disputes over water. Hillel asserts that the Six Day War between Israel and Syria in 1967 was related to a dispute over rights to water from the Jordan River. He also points to a 1990 dispute among Iraq, Syria, and Turkey over water from the Euphrates River and tensions between Saudi Arabia and Jordan over extractions from a thirty-thousand-year-old aquifer underlying the two countries, as evidence that water can contribute to, or cause, conflicts.

Several locations have been identified as places where climate change could spark a future conflict between nations over water resources. Places that are vulnerable to future water battles include ones where a country depends on water that is under the control of a different country, or where two countries with conflicting uses both declare rights to the same water source. The Nile River basin is one such location.

The waters of the Nile River and its two main tributaries, the White Nile and the Blue Nile, flow through several nations in northeast Africa before emptying into the Mediterranean Sea. Living within the Nile's basin are three hundred million people in ten countries: Burundi, Egypt, Eritrea, Ethiopia, Kenya, Rwanda, Sudan, Tanzania, Uganda, and the Democratic Republic of the Congo. Yet only Egypt and, to a lesser extent, Sudan, have rights to the vast majority of the Nile's water. This arrangement dates back to two agreements struck in 1929 and 1959, when the British were a colonial power in much of Africa. Under these agreements, upstream countries generally have no rights to the river's water without approval from Egypt and Sudan.

For many years upstream countries have been challenging Egypt's and Sudan's control over the water. Population growth and a lack of rainfall have increased demands for water in the region, particularly in Ethiopia. According to Water.org, less than 50 percent of the Ethiopian population has access to an improved water supply. Agriculture is at the mercy of seasonal rains, which are becoming increasingly erratic, resulting in more frequent droughts. Ethiopia is one of the poorest countries in Africa and a major recipient of humanitarian food aid.

The Ethiopian government asserts that its people would not need food aid if the country had a reliable source of water to irrigate its agricultural crops. Ethiopia would like to tap into the Nile River for irrigation water, but Egypt has been reluctant to grant approval for such projects.

Egypt is highly protective of the Nile, which is crucial to the country's survival. Egypt uses the river's water for drinking, to irrigate crops, for energy, and for industry. The entire Egyptian population depends on the Nile. In 1979 Egyptian President

*The Ethiopian government asserts that many Ethiopians face difficulties in meeting their needs for water due to limited access to the Nile River.* Getty Images.

Anwar Sadat said, "The only matter that could take Egypt to war again is water."[2] In 1988 then-Egyptian foreign minister Boutros Boutros-Ghali predicted that the next war in the Middle East would be fought over the waters of the Nile, not politics.[3] More recently, in 2010, the Egyptian minister of water resources and irrigation stated, "Nile water is a matter of national security to Egypt. We won't under any circumstances allow our water rights to be jeopardized."[4]

Ethiopia and other upstream countries claim that it is unfair for Egypt to control so much of the Nile's water, particularly since the majority of the river's water comes from upstream countries. In a 2005 story from BBC News, Ethiopia's prime minister Meles Zenawi was quoted as saying, "While Egypt is taking the Nile water to transform the Sahara Desert into something green, we in Ethiopia—who are the source of 85 percent of that water—are denied the possibility of using it to feed ourselves. And we are being forced to beg for food every year."[5] Meles and the leaders of other upstream countries are becoming increasingly angry with Egypt and they are threatening to build irrigation and hydroelectric plants, with or without Egypt's approval.

Climate change can only intensify the dispute over the Nile River. The populations of both Ethiopia and Egypt, as well as the entire region, are growing and placing more demands on the Nile. Hotter temperatures mean people, animals, and crops will need more water. At the same time, climate change is expected to cause a decrease in rainfall in much of the area—some say it already has. Thus, the Nile's water will become even more crucial for the region's survival. In 2010 Rwandan minister for environment and lands Stanislas Kamanzi commented on the vulnerability of his country to climate change and the need for the Nile River's water. According to Kamanzi, "Everyone knows we are all faced with the realities of climate change. Rwanda is one of the most vulnerable countries . . . 80 percent of our agriculture relies on rain-fed irrigation . . . and we can no longer predict that we will receive regular rainfall, so we can't predict our crop production. We need to use the resources in our waterways and lakes, and these are clearly in the Nile Basin."[6] Unless a diplomatic solution is reached, climate change may provide the spark to ignite a conflict over the Nile River.

## Fighting over Land

Not far from Uganda and Ethiopia sits Darfur, Sudan, a place where many people think a climate-change-fueled conflict has already occurred. Darfur lies between two deserts and has little land that can sustain life. What land is available for farming or livestock grazing is fragile semiarid land vulnerable to desertification. Desertification occurs when semiarid lands are turned into arid lands, or deserts, though persistent droughts and human exploitation. In Darfur, pastoralists—nomadic people who depend on livestock for survival—have overgrazed the land, while farmers have overcultivated it. Additionally, in the past several years, the region has experienced frequent droughts. The semiarid land in Darfur is fast becoming a desert.

*"Sudan's tragedy is not just the tragedy of one country in Africa—it is a window to a wider world underlining how issues such as uncontrolled depletion of natural resources like soils and forests allied to impacts like climate change can destabilize communities, even entire nations."*

In 2003 rising tensions between the pastoralists, who are largely Arab, and the farmers, who are largely African, ignited into widespread violence. In addition to competition for scarce land, the violence has been blamed on ethnic, cultural, and religious differences. The conflict has taken a terrible toll on the region. Hundreds of villages were destroyed, as many as three hundred thousand people died because of violence or starvation, and millions of people have been forced to leave their homes. Humanitarian groups, as well as the US government, have said that the events in Darfur constituted genocide. They assert that Arab militias—with the help of the Sudanese government— tried to wipe out the African populace. In 2010 the International Criminal Court agreed and charged the Sudanese president, Omar al-Bashir, with genocide.

In 2007 the United Nations Environment Program (UNEP) released a report examining the causes of the Darfur conflict. The report cited climate change as being among the main causes of the crisis. According to the report, millions of hectares (acres) of land in the Darfur region have been lost to desertification. UNEP says that a marked decline in rainfall—in northern Darfur precipitation has fallen by a third since the early 1900s—clearly indicates climate change is occurring in the region on an "almost unprecedented scale."[7] At the release of the report, Achim Steiner, UNEP's executive director, warned that tragedies like Darfur could become more common with climate change. According to Steiner, "Sudan's tragedy is not just the tragedy of one country in Africa—it is a window to a wider world underlining how issues such as uncontrolled depletion of natural resources like soils and forests allied to impacts like climate change can destabilize communities, even entire nations."[8]

## The Link Between Climate Change, Migration, and Conflict

Darfur also highlights what many experts believe will be one of the most significant impacts of climate change on humans, and one that is also expected to lead to conflict: migration. In 2007, researchers at the International Peace Research Institute in Oslo, Norway, published a paper examining climate change and conflict. According to the researchers, Nils Petter Gleditsch, Ragnhild Nordås, and Idean Salehyan, "migration is one of the most plausible links from climate change to conflict."[9]

Gleditsch, Nordås, and Salehyan assert that there are two pathways whereby climate change can cause migration. First, climate change may create environmental refugees, people who mi-

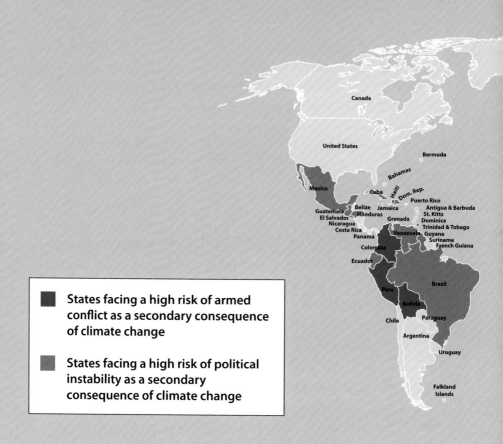

States facing a high risk of armed conflict as a secondary consequence of climate change

States facing a high risk of political instability as a secondary consequence of climate change

Source: Dan Smith and Janani Vivekanandapp, *A Climate of Conflict*, International Alert, November 2007. www.international-alert.org/pdf/A_Climate_Of_Conflict.pdf.

grate directly because an environmental stress—such as drought, flood, or sea-level rise—makes an area uninhabitable. Second, climate change may create conflict refugees. In this case, environmental stress caused by climate change may result in conflicts over scarce resources, and the conflict, rather than the environmental stress, may be the direct cause of the migration. Whatever

the pathway, climate change may result in migration, which in turn can lead to conflict and violence.

Migration may lead to conflict in a number of ways. Tensions, and conflict, can arise if the people in the receiving area feel threatened by the new migrants. For instance, locals may feel the migrants are detracting from their ability to find a job or make a living. They may also be threatened by the new arrivals' consumption of resources, such as water, food, or land. If pressures get high, people may attempt to secure these resources by force. Ethnic, cultural, or religious differences between the migrants and the people in the receiving area may also spark conflict, especially if these divisions are long-standing. Finally, migrants fleeing from conflict may bring violence to the receiving area with them.

The states of northeastern India at the northern border of Bangladesh may be a region prone to conflicts caused by climate-change-induced migration. As described in Chapter 5, Bangladesh is a climate-change hot spot. The country is vulnerable to sea-level rise and flooding. Bangladeshis' standard of living is very poor, and their ability to adapt to climate change is limited. Additionally, the population of this already densely populated country is growing. Experts predict these factors may cause thousands, or even millions, of environmental refugees to flee from Bangladesh into neighboring India.

History indicates that such an influx of migrants from Bangladesh into India could lead to conflicts. For instance, in the 1980s, tensions ignited into violence in the Indian states of Assam and Tripura over Bangladesh migrants. In Assam, violence erupted when the natives accused the newcomers of stealing their lands. In the state of Tripura, the newly arrived Bangladeshis came in such numbers that they overwhelmed the native population and, over several years, became the majority. Resentful and increasingly competitive with the migrants over land and resources, the native people turned to violence. The conflicts in Assam and Tripura illustrate what could happen in the future if climate change creates large numbers of environmental refugees from Bangladesh.

The people from Bangladesh would be considered environmental refugees, but climate change may also create conflict refugees, as described by Gleditsch, Nordås, and Salehyan. The conflict in Darfur provides an example of how climate change may create this type of refugee. As described above, dwindling land, due in part to climate change, helped ignite conflict in Darfur. As a result of the conflict, millions were displaced from their homes. According to the organization Amnesty International, most of those driven from their homes ended up in camps for internally displaced people (IDPs) on the outskirts of Darfur. IDPs are people who have been displaced from their homes but still reside in the country of their origin. In addition to camps, other IDPs squatted in shacks, lived with relatives, or just sheltered in the bush. Others escaped to towns and cities elsewhere in Sudan. About a quarter of a million people from Darfur migrated to refugee camps in Chad. They are classified as international refugees because they were driven across the border into a different country.

Not long after the Darfur refugees arrived in Chad, conflict arose there too. In 2006 news media reported numerous incidents in which the Arab militia followed fleeing Africans into Chad and began attacking the refugees and the Chadians living near the refugee camps. In an interview with National Public Radio, *New York Times* columnist Nicholas Kristof expressed his fear that the violence was going to keep spreading, as he recounted the devastating experience of a young woman from Darfur. According to Kristof:

> I talked to one young woman on this trip, who initially was targeted in March [2006] and she was gang-raped by the Janjaweed [Arab militia], at that time, and her 10-year-old sister was raped and then killed. This young woman—Hilema...—then fled further inland into Chad where she thought she was going to be safe. But now the Janjaweed have followed her even into that area and last month she was seized again and gang-raped again. You just have the feeling that this is going to go on and, you know, maybe topple all of Chad and all of Central African Republic.[10]

## Climate Change Is a Threat Multiplier

Although Chad, Darfur, India, and Bangladesh are far away from North America, US military strategists are also concerned

## US Congressional Testimony on the Potential for Conflict in the Arctic

Ice in the Arctic Ocean is melting much more quickly than most people appreciate and US policymaking is lagging far behind environmental realities. *The Arctic is the fastest warming region on earth and is on pace to be ice free in the summer by 2013.* The past few years have witnessed extraordinary melting and last summer [2008] the two fabled Arctic passages over Eurasia and North America opened together for the first time in history. Recent satellite images of the Chuchki and Beaufort Seas show dramatically less ice than what is historically normal for this time of year. By every measure, from huge ice shelves breaking free to complex environmental dynamics that scientists do not fully understand, the polar ice cap is disappearing and all indicators point to another record sea ice minimum this coming summer. We may be approaching a tipping point past which the melting sea ice cannot recover. . . .

The Arctic is home to an estimated twenty-two percent of the world's remaining undiscovered hydrocarbon reserves as well as ac-

about climate change and conflict. In 2007 the nonprofit CNA Corporation's military advisory board—made up of eleven retired generals and admirals—issued a report examining the national security consequences of climate change. Among the military experts' findings was that climate change will act as a threat multiplier for instability in volatile and fragile regions of the world. The report contends that water scarcity, declines in food production, increases in diseases, and eroding economic and environmental conditions will cause large populations to move in search of resources. In addition, the report asserts, "weakened and failing governments, with an already thin margin for survival, foster the conditions for internal conflicts, extremism, and movement toward increased authoritarianism and radical ideologies."[11] The retired military officers go on to state that even the United States

cess to the fabled shipping routes over Eurasia and North America, both of which have led to balance-of-power struggles in the region. The next few years will be critical in determining whether the Arctic's long-term future will be one of international harmony and the rule of law, or of a Hobbesian free-for-all with dangerous potential for conflict.

This is a story still being written with a plot full of characters who speak of multilateral cooperation but pursue their own self-interest. . . . Every single bilateral relationship where Arctic countries share a physical border, except one, Norway and Denmark, has at least one significant point of disagreement. Like previous assumptions that the icecap is melting more slowly than it actually is, it would be a mistake to assume that all these potential flashpoints will remain sleeping dogs. The combination of new shipping routes, trillions of dollars in possible oil and gas resources, and a poorly defined picture of state ownership make for a toxic brew.

SOURCE: **Statement of Scott G. Borgerson, Council on Foreign Relations, Before the US House of Representatives Committee on Foreign Affairs, Washington, DC, March 25, 2009.**

and other stable regions of the world may experience tensions because of climate change. For instance, the United States and Europe may experience mounting pressure to accept large numbers of immigrant and refugee populations from Latin America and Africa. Additionally, extreme weather events and natural disasters, like what the United States experienced with Hurricane Katrina, may lead to increased missions for an already stretched military. Among the retired generals' and admirals' proposals to avoid climate change conflicts was the recommendation that the United States commit to global partnerships that help developing nations build the capacity and resiliency to better manage climate impacts.

# Notes

1. Paul Simon, *Tapped Out: The Coming World Crisis in Water and What We Can Do About It*. New York: Welcome Rain, 2002, p. 4.
2. Quoted in Patricia Kameri-Mbote, "Water, Conflict, and Cooperation: Lessons from the Nile River Basin," N*avigating Peace Initiative: Water Stories*. Washington, DC: Woodrow Wilson International Center for Scholars, January 2007.
3. Kameri-Mbote, "Water, Conflict, and Cooperation: Lessons from the Nile River Basin."
4. Amro Hassan, "Egypt: Minister Rejects Nile Sharing Deal as Experts Warn of Water Shortage," *Los Angeles Times*, April 20, 2010.
5. Mike Thomson, "Nile Restrictions Anger Ethiopia," *BBC News*, February 3, 2005.
6. Quoted in Mike Pflanz, "Egypt, Sudan Lock Horns with Lower Africa over Control of Nile River," *Christian Science Monitor*, June 4, 2010.
7. United Nations Environment Program, *Sudan: Post Conflict Environmental Assessment*, 2007.
8. UN News Center, "UN Report Says Environmental Degradation Triggering Tensions in Sudan," June 22, 2007.
9. Nils Petter Gleditsch, Ragnhild Nordås, and Idean Salehyan, *Climate Change and Conflict: The Migration Link*. New York: International Peace Academy, May 2007.
10. Melissa Block, "Violence Spills from Darfur into Chad," *NPR*, November 16, 2006.
11. CNA Corporation, *National Security and the Threat of Climate Change*, 2007.

CHAPTER 7

# Conclusion: Reducing the Human Imprint on the Natural World

Climate change has caused an increased awareness around the world of humanity's imprint on the earth. The Intergovernmental Panel on Climate Change Fourth Assessment Report, issued in 2007, laid most of the blame for climate change on human-generated greenhouse gas emissions. In an attempt to mitigate and adapt to a warmed planet, many people are calling for policies that slow population growth and that promote environmentally friendly economic development.

Some scientists think that population policies can help to mitigate climate change and reduce its impact on the human population. As discussed in Chapter 3, the growing human population and our increasing consumption of resources, particularly energy, are inextricably linked to climate-warming greenhouse gas emissions. Some scientists contend that this connection provides a compelling reason to slow population growth and stabilize the human population. In a 2008 paper, British physicist Martin Desvaux and reproductive health specialist John Guillebaud argue that policy makers must slow population growth in order to solve climate change. They argue that climate change is just one example of humanity's environmental impact, and one that is caused in large part by the sheer number of humans. As the population is multiplying, they say, so are the number of hungry people in the world. In their paper, Guillebaud and Desvaux state the following:

We have lost sight of the fact that Earth has limits. When species multiply beyond the capacity of their environment, nature provides no alternative to a die off. Unprecedented global disasters loom and *all of us* are the problem: The rich, because of how they wastefully over consume (that *must change*), and the poor, because of their reasonable aspirations to leave poverty, which means increased consumption and hence inevitably more greenhouse gas production per person. And the number of persons in *both* subsets of the population is steadily increasing.[1]

Desvaux and Guillebaud believe the world needs an urgent plan to curb overpopulation. Removing the barriers to family planning worldwide, including increasing access to contraceptives and providing birth-control information, must be part of that plan, they argue.

---

*According to the nonprofit Population Reference Bureau, in 2008 women in developed countries had 1.6 children on average, whereas women in developing countries had 2.9 children on average.*

---

## The Role of Population Policies in Mitigation

In developing countries, women tend to have more children than they desire, generally because of a lack of family planning services and modern contraception. According to the nonprofit Population Reference Bureau, in 2008 women in developed countries had 1.6 children on average, whereas women in developing countries had 2.9 children on average. In 2009, the Guttmacher Institute and the United Nations Population Fund (UNFPA) issued a report on the developing world's unmet need for family planning. Using data from 2008, the report estimated that as many as 215 million women in developing

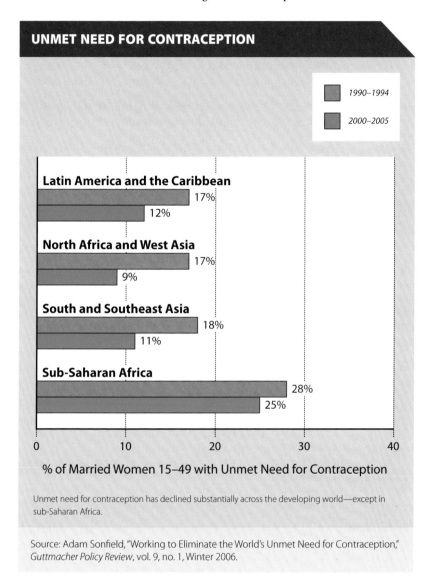

## UNMET NEED FOR CONTRACEPTION

1990–1994

2000–2005

**Latin America and the Caribbean**
17%
12%

**North Africa and West Asia**
17%
9%

**South and Southeast Asia**
18%
11%

**Sub-Saharan Africa**
28%
25%

% of Married Women 15–49 with Unmet Need for Contraception

Unmet need for contraception has declined substantially across the developing world—except in sub-Saharan Africa.

Source: Adam Sonfield, "Working to Eliminate the World's Unmet Need for Contraception," *Guttmacher Policy Review*, vol. 9, no. 1, Winter 2006.

countries want to avoid pregnancy but are not using an effective means of contraception, such as birth control pills or injections, IUDs (intrauterine devices), or male and female sterilization. Young, poor, uneducated, rural women in low-income countries, such as ones in sub-Saharan Africa, Pakistan, and

Bangladesh, are more likely than other women to have an unintended pregnancy.

The Guttmacher Institute and UNFPA say that if all women in the developing world had access to modern contraception, the number of unintended pregnancies would drop significantly. If the 215 million women with unmet need had used modern family planning methods, unintended pregnancies would drop by 71 percent, from 75 million to 22 million per year, they say.

Desvaux, Guillebaud, and others believe family planning programs that slow population growth can help to mitigate climate change. Desvaux and Guillebaud believe that family planning programs may not be the only remedy to blunt climate change, but "it's an utterly crucial and tragically underfunded one."[2] In 2008 demographer and ecologist Frederick Meyerson called for more US funding of family planning programs in the developing world. According to Meyerson, "If the United States were to increase its assistance for population programs by $1 billion annually, and other donor countries contributed their share, it should be possible to satisfy the global unmet need for family planning within five years. As a result, the population growth rate could be reduced by about 30 percent, with a similar decrease in the growth of greenhouse gas emissions."[3]

Not all scientists think that reducing unintended pregnancies will have much impact on curbing global greenhouse gas emissions. Joseph Chamie, demographer and editor of the journal *International Migration Review*, believes that the overconsumption of resources by more developed countries is the real culprit in climate change, not the growing population of the developing world. According to Chamie, "helping women and men who want to avoid or delay pregnancy is a laudable goal," but, "focusing on reducing unintended fertility to address climate change—in particular to decrease global greenhouse gas emissions—strikes me as a delay tactic. Instead, the focus should be on significantly and immediately reducing damaging patterns of production and consumption. That's where we can make the

real difference."[4] Associate professor, reproductive rights specialist, and women's health advocate Betsy Hartmann agrees. She
rails against the Malthusian premise, described in more detail in
Chapter 2, that equates population growth with environmental
degradation and poverty. According to Hartmann:

> There is a profound difference in the worldviews of those who
> see the roots of poverty and environmental degradation in
> overpopulation, and those who locate them in structural eco-
> nomic, political, and social (including gender) inequalities. In
> Bangladesh, where I lived in the 1970s, villagers were poor not
> because they had too many children, but because centuries of
> colonial rule and corrupt governance concentrated land, re-
> sources, and power in the hands of a few.

Hartmann goes on to say that, "after witnessing such in-
equalities, I could never again accept the simplistic—and often
elitist—assumptions of Malthusian thought."[5] Hartmann and
Chamie think that it is an oversimplification to look to family
planning programs as a means to mitigate climate change.

## The Role of Population Policies in Adaptation

Policy makers in developing countries also believe family planning programs can play an important role in climate change, but
not as a tool to decrease greenhouse gas emissions. They see family planning programs as a way to reduce vulnerability and help
their countries adapt to climate change impacts. Under the United
Nations Framework Convention on Climate Change, developing countries and small island states that develop climate change
adaptation programs—known formally as National Adaptation
Programs of Action (NAPAs)—can receive financial assistance
from wealthier countries. In 2009 researchers in England,
Australia, and Ethiopia studied the NAPAs of forty developing countries, including Bangladesh, Ethiopia, Rwanda, Sudan,
Tuvalu, Uganda, Tanzania, and Yemen. The researchers—Leo

*Women drag sacks of food provided by relief agencies in Uganda. Under climate change, population growth in Uganda may further stress the nation's resources.* Walter Astrada/AFP/ Getty Images.

Bryant, Louise Carver, Colin Butler, and Ababu Anaged—found that many of the NAPAs they studied identified rapid population growth as a problem that either exacerbates the effects of climate change or impedes the ability to adapt to climate change. Bryant, Carver, Butler, and Anaged state that while the concerns of the different NAPA reports are diverse, three key themes emerge. First, both population growth and climate change can cause a reduction in the supply of natural resources. Second, rapid population growth is projected to escalate the demand for resources that are diminished by climate change, including freshwater and food. Third, "rapid population growth heightens human vulnerability to natural disasters caused by climate change, such as by forcing more people to migrate and settle in areas at risk of floods, storms, drought and infectious disease."[6]

According to the Uganda NAPA, which was issued in 2007, under a climate change scenario the country's already stressed resources will not be able to supply enough land, water, energy, food, or other resources to keep up with rapid population growth. The NAPA identifies controlling overpopulation with family planning programs as a necessary activity in responding to climate change. Similarly, Ethiopia's NAPA, also issued in 2007, identifies family planning as an adaptation option.

Bryant, Carver, Butler, and Anaged recommend that voluntary family planning services be made more available to poor communities in least-developed countries to assist their ability to adapt to the harmful effects of climate change. Many nonprofit environmental and social organizations agree. The organization Population Action International says that, "slower population growth will reduce the scale of vulnerability to the effects of climate change, and make reductions in global greenhouse gas emissions easier to achieve. Increasing access to family planning is a desired strategy to help people adapt to the inevitable effects of climate change."[7] In their *State of the World 2009*, the Worldwatch Institute says it is essential to "promote policies and programs that can help slow and eventually reverse [population] growth by making sure that all women are able to decide for themselves whether and when to have children."[8]

## Empowering Women

Family planning programs may also help with climate change mitigation and adaptation through the positive impact they have on women. Family planning can improve, and even save, women's lives. According to the Population Reference Bureau (PRB), half a million women, mostly in the developing world, die each year from pregnancy-related causes. The PRB says this number could be reduced by at least one third if all the women who wanted to postpone their next pregnancy or stop having children altogether had access to modern contraception. Beyond the positive health effects, family planning can provide other benefits for the lives of women. The PRB claims that "when women delay their next birth or have fewer children, the potential to educate, train, and meet the economic demands of a young population becomes

## The Sinsibere Project in Mali: Women as Environmental Stewards

Mali is a Sahelian country of which two thirds is desert. Ninety per cent of the country's energy needs are met by burning wood and charcoal. As a result, deforestation is intensifying and desertification is accelerating. Loss of wood cover is intensifying erosion, which in turn makes the soil poorer for farming and exposes loose soil that is more vulnerable to flood. Flooding happens more often with the heavy rains, and this is seen as partly due to climate change. The Sinsibere project works to reduce desertification by developing sustainable sources of income for rural women as an alternative to their commerce in wood. These alternative livelihoods include vegetable gardens and making shea butter products like soap. After six years, 80 percent of the participating women no longer cut wood for commercial purposes, or [they] have substantially reduced their wood-cutting. Besides environmental, literacy and financial education, training in soap making and in making energy-efficient stoves was organized for the rural women. These different trainings provided the women with skills that have made them more confident about themselves, better able to explore alternative livelihood options, and more eager to participate in village decision-making.

SOURCE: United Nations International Strategy for Disaster Reduction, *Gender Perspectives: Integrating Disaster Risk Reduction into Climate Change Adaptation: Good Practices and Lessons Learned*, 2008.

easier. Slower population growth can yield savings on the costs of providing health, clean water, sanitation, and social services. Meeting women's desires to avoid or postpone pregnancies also allows women more opportunities for education and employment, which can spur economic growth."[9]

According to UNFPA and the Women's Environment & Development Organization, "Women are not only victims of climate change, but also effective agents of change in relation to both mitigation and adaptation. Women have a strong body of knowledge and expertise that can be used in climate change mitigation, disaster reduction and adaptation strategies."[10] In the view of Thoraya Ahmed Obaid, executive director of UNFPA, "women have the power to mobilize against climate change, but this potential can be realized only through policies that empower them."[11]

## Sustainable Development

Economic growth is an important factor in raising the living standard of developing countries and lifting people out of poverty. According to the United States Agency for International Development (USAID), "economic growth is the surest way for countries to generate the resources they need to weather global crises—from unstable markets for finance to those for energy and food—and to address their own illiteracy, poor health and other long-term development challenges."[12]

Many scholars and commentators emphasize the need for economic growth in developing countries, as well as in developed countries, to occur in a sustainable fashion. Generally, economic growth, or economic development, leads to increases in greenhouse gas emissions. This increase can be minimized when development occurs sustainably, however.

Sustainable development refers to a development approach that allows current generations to meet their economic and social needs while also preserving the earth's resources and the environment so that future generations can meet *their* economic and social needs. Sustainable development can take many different forms. For instance, sustainable forest management refers to timber harvests that leave enough timber in place for future generations. Sustainable development also applies to building construction, specifically the use of recycled materials, efficient water and energy use, and a smaller carbon footprint.

Clean, efficient, and renewable energy is an important component of sustainable development. According to the United

Nations, worldwide, approximately 3 billion people rely on traditional biomass—wood, agricultural waste, or dung—for cooking and heating, and about 1.5 billion have no access to electricity. Up to a billion more have access only to unreliable electricity networks. Developing the infrastructure and projects to provide electricity to people without it is a key part of economic development and social progress. Schools, health clinics, water and sanitation systems, and various means of food preservation all depend in some way on the availability of a modern electricity supply. Yet the generation of electricity is a major contributor to climate change. According to the United Nations, the energy system—supply, transformation, delivery, and use—represents about 60 percent of total current (2010) greenhouse gas emissions. Ensuring that new electricity projects use renewable fuels, such as wind, solar, or hydroelectricity, and that appliances are energy efficient, will allow developing countries to grow economically with the smallest possible carbon footprint.

Although considerable obstacles exist, it may be possible to help people achieve a higher standard of living and simultaneously address climate change. Population policies, such as expanded access to reproductive health and family planning services, may reduce vulnerability to climate change impacts and slow the growth of greenhouse gas emissions, while also leading to greater educational and economic opportunities for women. Sustainable development can provide energy and economic growth to people now without compromising future generations. According to Helen Clark, administrator of the United Nations Development Program, "Mitigating and adapting to climate change is entirely compatible with pursuing development."[13] The possibility exists for humanity to reduce its imprint on the earth and to adapt, and even thrive, in a future warmed world.

## Notes

1. Martin Desvaux and John Guillebaud, "Curbing Population Must Contribute to Solving the Climate Crisis," *Bulletin of the Atomic Scientists Population and Climate Change Roundtable*, February 7, 2008.

2. Desvaux and Guillebaud, "Curbing Population Must Contribute to Solving the Climate Crisis."

3. Frederick Meyerson, "Reducing Unintended Fertility Should Be a Top International Climate Priority," *Bulletin of the Atomic Scientists Population and Climate Change Roundtable*, February 15, 2008.

4. Joseph Chamie, "Reducing Unintended Fertility Will Have Little Impact on Emissions," *Bulletin of the Atomic Scientists Population and Climate Change Roundtable*, February 22, 2008.

5. Betsy Hartmann, "Sustainable Lifestyles, Not Population Control, Will Solve the Climate Crisis," *Bulletin of the Atomic Scientists Population and Climate Change Roundtable*, January 29, 2008.

6. Leo Bryant, Louise Carver, Colin Butler, and Ababu Anaged, "Climate Change and Family Planning: Least-Developed Countries," *Bulletin of the World Health Organization*, 2009.

7. Population Action International, *Meeting the Development and Health Needs of 215 Million Women: U.S. International Family Planning Goals*, 2008.

8. Christopher Flavin and Robert Engelman, "The Perfect Storm," *State of the World 2009: Worldwatch Institute*, 2009.

9. Mary Mederios Kent, "What Would It Cost to Meet Family Planning Needs in Developing Countries?" Population Reference Bureau, 2010.

10. United Nations, "Gender Perspectives on Climate Change," Fifty-Second Session of the Commission on the Status of Women, February 28, 2008.

11. United Nations Population Fund, *State of World Population 2009: Facing a Changing World: Women, Population and Climate*, 2010.

12. United States Agency for International Development, "Economic Growth and Trade," www.usaid.gov.

13. United Nations Development Program, *Charting A New Low-Carbon Route to Development*, June 2009.

# Glossary

**adaptation** An adjustment in natural or human systems in response to actual or expected climatic stimuli or their effects; moderates harm or exploits beneficial opportunities.

**adaptive capacity** The ability of a system to adjust to climate change, to moderate potential damage, to take advantage of opportunities, or to cope with the consequences of such an altered environment.

**anthropogenic** Produced by human beings or resulting from human activity.

**arid** A region lacking moisture, especially one having insufficient rainfall to support trees or woody plants.

**bottleneck** An evolutionary event in which a significant percentage of a population or species is killed or otherwise prevented from reproducing.

**carbon cycle** The term used to describe the flow of carbon (in various forms, e.g., carbon dioxide) through the atmosphere, ocean, terrestrial biosphere, and the lithosphere (crust and upper mantle of the earth).

**carbon dioxide ($CO_2$)** A naturally occurring gas fixed by photosynthesis into organic matter. A by-product of fossil-fuel combustion and biomass burning, it is also emitted from land-use changes and other industrial processes. It is the principal anthropogenic greenhouse gas.

**carrying capacity** The maximum sustainable size of a resident population in a given ecosystem.

**cholera** A severe intestinal infection caused by the ingestion of contaminated water or food.

**combustion** The rapid oxidation of a fuel to release heat energy, or, burning.

**contraception** Intentional prevention of conception or impregnation through the use of various devices, agents, drugs, sexual practices, or other methods.

**deforestation** A natural or anthropogenic process in which forest land is converted to nonforest land.

**demographic transition** The historical shift of birth and death rates from high to low levels in a population. A decline of mortality rates usually precedes a decline in fertility rates, thus resulting in rapid population growth during the transition period.

**demography** The branch of science that studies the statistics and characteristics of human populations.

**desertification** The gradual transformation of habitable land into desert.

**drought** An extended period in which precipitation is significantly below normal recorded levels, causing serious hydrological imbalances that often adversely affect land resources and production systems.

**emigration** The process of leaving one country to take up permanent or semipermanent residence in another.

**emission** A substance discharged into the air or water.

**evaporation** The transition process from liquid to gaseous state.

**extinction** The global disappearance of an entire species.

**family planning** The conscious effort of couples to regulate the number and spacing of births through artificial and natural methods of contraception.

**fertility** The reproductive performance of an individual, a couple, a group, or a population.

**fertilization effect** The stimulation of plant photosynthesis through elevated carbon dioxide concentrations, leading to enhanced productivity.

**food security** A situation that exists when people have secure access to sufficient and safe nutrition for normal growth, development, and an active and healthy life.

**fossil fuels** Coal, petroleum (oil), and natural gas.

**glacier** A slowly moving mass of ice that flows over land.

**globalization** The growing integration and interdependence of countries worldwide through the increasing volume and variety of cross-border transactions in goods and services; free international capital flows; and the more rapid and widespread diffusion of technology, information, and culture.

**gross domestic product (GDP)** The monetary value of all goods and services produced within a nation.

**gross national product (GNP)** The monetary value of all goods and services produced in a nation's economy, including income generated abroad by domestic residents, but without income generated by foreigners.

**groundwater** Water that exists beneath the earth's surface in underground streams and aquifers. Groundwater flows naturally to the earth's surface via seeps or springs, or it can be extracted through wells, tunnels, or drainage galleries.

**groundwater recharge** The replenishment of groundwater supplies from rain or overland flow.

**habitat** The locality or natural home in which a particular plant, animal, or group of closely associated organisms lives.

**hydrological cycle** The continuous movement of water on, above, and below the surface of the earth.

**ice age** A geological period characterized by reduced temperatures on the earth's surface and in the atmosphere, resulting in an expansion of ice sheets and glaciers.

**ice sheet** A mass of land ice that is sufficiently deep to cover most of the underlying bedrock topography. Only two large ice sheets exist in the modern world—on Greenland and on Antarctica.

**immigration** The process of entering one country from another to take up permanent or semipermanent residence.

**infiltration** The process by which water on the ground surface enters the soil.

**infrastructure** The basic equipment, utilities (e.g., electricity or water), productive enterprises, installations, and services essential for the development, operation, and growth of a city or nation.

**internally displaced people (IDPs)** Individuals who are forced to flee their homes but who remain within their country's borders.

**irrigation** The process of supplying water to drylands for agriculture or landscaping through artificial processes, such as the building of dikes.

**Malthusian** Of or relating to economist Thomas R. Malthus (1766–1834) and his theory that the world's population tends to increase faster than the food supply and that unless fertility is controlled, famine, disease, and war must serve as natural population restrictions.

**megacity** Generally defined as a metropolitan area with a population in excess of 10 million people.

**Millennium Development Goals (MDGs)** A list of ten aims, including eradicating extreme poverty and hunger, improving maternal health, and ensuring environmental sustainability, adopted in 2000 by the United Nations General Assembly to be reached by 2015.

**morbidity rate** Frequency of injury, sickness, or disease.

**mortality rate** Frequency of death.

**neo-Malthusian** A modern advocate of Thomas Malthus's population theory.

**nomad** A member of a group of people who have no fixed home and move from place to place, according to the seasons, in search of food, water, and grazing land for livestock.

**pastoralist** A person involved in pastoral farming; someone whose primary occupation is the raising of livestock.

**per capita** Per person.

**permafrost** Perennially frozen ground that occurs where the temperature remains below 0°C for several years.

**photosynthesis** The process whereby plants produce sugars from sunlight and carbon dioxide and produce oxygen as a by-product.

**rangeland** Unmanaged grasslands, scrublands, savannas, and tundra.

**refugee** Defined by the United Nations Convention Relating to the Status of Refugees as a person who, "owing to a well-founded fear of being persecuted for reasons of race, religion, nationality, membership of a particular social group, or political opinion, is outside the country of his nationality, and is unable to or, owing to such fear, is unwilling to avail himself of the protection of that country."

**resilience**  Refers to three conditions that enable social or ecological systems to bounce back after a shock, such as from climate change. The conditions are an ability to self-organize, an ability to buffer disturbance, and a capacity for learning and adapting.

**respiration**  The cellular process whereby living organisms obtain energy by converting organic matter and oxygen into carbon dioxide.

**runoff**  Rainfall not absorbed by soil.

**Sahel**  A semiarid region south of the Sahara Desert stretching from Senegal to Egypt.

**salinization**  The accumulation of salts in soils.

**saltwater intrusion**  Displacement of fresh surface water or groundwater by the advance of saltwater.

**savanna**  Tropical or subtropical grassland or woodland biomes scattered with shrubs and trees, and characterized by a dry climate.

**semiarid**  Regions of moderately low rainfall that are not highly productive and are usually classified as rangelands.

**slash and burn**  A agricultural technique wherein trees and other plants are cut down and then incinerated to prepare fields for the next crop.

**Southeast Asia**  A region consisting of the countries that are geographically south of China, east of India, and north of Australia.

**sub-Saharan Africa**  The region of Africa south of the Sahara Desert.

**subsidence**  A gradual sinking of land relative to its previous level.

**surface runoff** Water that travels over the land surface to the nearest surface stream.

**sustainable** Using natural and human resources in a way that does not jeopardize the opportunities of future generations.

**sustainable development** Development that meets the cultural, social, political, and economic needs of the present generation without compromising the ability of future generations to meet their own needs.

**temperate** Geographic regions with moderate temperatures and climate.

**tsunami** A large ocean wave caused by an underwater earthquake or volcanic eruption.

**tundra** A treeless plain—level or gently undulating—typical of the Arctic and sub-Arctic regions, characterized by low temperatures and short growing seasons.

**urbanization** The movement of people from rural to urban areas.

**vulnerability** The degree to which a system is susceptible to, and unable to cope with, adverse effects of climate change, including climate variability and extremes.

# For Further Research

## Books

Lael Brainard, Abigail Jones, and Nigel Purvis, *Climate Change and Global Poverty: A Billion Lives in the Balance?* Washington, DC: Brookings Institution Press, 2009.

> Brainard, Jones, and Purvis contend that poverty alleviation must become a central strategy for reducing global vulnerability to adverse climate impacts. They discuss how climate change solutions can empower global development by improving livelihoods, health, and economic prospects.

William James Burroughs, *Climate Change in Prehistory: The End of the Reign of Chaos*. New York: Cambridge University Press, 2005.

> The late writer and scientist Burroughs weaves together studies of the climate with anthropological, archaeological, and historical studies to explain how climate change in prehistory has essentially made us who we are today. Burroughs reviews how certain aspects of human physiology and intellectual development—as well as our language, diets, and domestication of animals—have all been influenced by climatic factors occurring far back in human history.

Matthew Connelly, *Fatal Misconception: The Struggle to Control World Population*. Cambridge, MA: Harvard University Press, 2008.

> Columbia University professor Connelly recounts the global history of the population control movement, including discussions of eugenics, sterilization camps in India, and China's one-child policy. Connelly contends that despite their sometimes good intentions, family planning programs have caused untold suffering.

Andrew Goudie, *The Human Impact on the Natural Environment: Past, Present, and Future*. Hoboken, NJ: Wiley-Blackwell, 2005.

> Goudie provides a comprehensive view of the major environmental issues facing the world today and explores the impact of humans on vegetation, animals, soils, water, landforms, and the atmosphere.

Stephen Humphreys, *Human Rights and Climate Change.* New York: Cambridge University Press, 2010.

> Humphreys provides a collection of articles exploring the human rights implications of climate change. Humphreys and the contributing authors discuss the difficulties of crafting government policies to deal with such issues as forced mass migration; increased disease incidence; threatened food and water security; the disappearance and degradation of shelter, land, livelihoods, and cultures; and the threat of conflict.

James R. Lee, *Climate Change and Armed Conflict: Hot and Cold Wars.* New York: Routledge, 2009.

> Lee, who runs American University's Inventory of Conflict and Environment project, examines the relationship between climate change and conflict. Lee argues that climate change will produce (as history has already demonstrated) two types of conflict: cold wars and hot wars.

Jeremy Lind and Kathryn Sturman, eds., *Scarcity and Surfeit: The Ecology of Africa's Conflicts.* Pretoria, South Africa: Institute for Security Studies and African Centre for Technology Studies, 2002. www.acts.or.ke.

> Examines two volatile regions of sub-Saharan Africa (the Greater Horn of Africa and the Great Lakes) analyzing the role of ecological factors, including natural resources, in regional conflicts.

Robin Mearns and Andrew Norton, *Social Dimensions of Climate Change: Equity and Vulnerability in a Warming World.* Washington, DC: World Bank, 2010.

> Mearns and Norton have compiled many of the papers presented at a 2008 international workshop discussing the challenges related to reducing poverty while also mitigating climate change. The book focuses on the social dimensions of climate change, in which policies meant to counteract climate change could undermine poverty reduction programs and reverse recent economic gains in developing countries.

National Research Council, *Beyond Six Billion.* Washington, DC: National Academies Press, 2000.

> The premier US scientific group, the National Research Council (NRC), analyzes population growth forecasts. The authors examine what population forecasts really say, why they say it, whether they

can be trusted, and whether they can be improved. The NRC explores fertility estimates, predictions of lengthening life span, and the impacts of international migration, and it discusses whether developing countries will ever attain the very low levels of births seen in the industrialized world.

Brian C. O'Neill, F. Landis MacKellar, and Wolfgang Lutz, *Population and Climate Change*. Cambridge, MA: Harvard University Press, 2001.
> A systematic, in-depth examination of the links between population growth and climate change. A multidisciplinary team of authors provides both natural science and social science perspectives.

Gavin Schmidt and Joshua Wolfe, *Climate Change: Picturing the Science*. New York: Norton, 2009.
> Schmidt, a climate scientist at NASA, and Wolfe, a photographer, provide an easy-to-understand explanation of climate science and its many potential impacts upon human lives, along with stunning images of climate impacts. The authors address the melting of ice and permafrost at the poles, the impact upon cities such as Venice and Miami of rising sea levels, and a host of other climate change impacts.

## Periodicals and Internet Sources

Sharon Begley, "The Evolution of An Eco-Prophet," *Newsweek*, October 31, 2009.

David Biello, "Can Climate Change Cause Conflict? Recent History Suggests So," *Scientific American*, November 23, 2009.

Matthew Bunce, Katrina Brown, and Sergio Rosendo, "Policy Misfits, Climate Change and Cross-Scale Vulnerability in Coastal Africa: How Development Projects Undermine Resilience," *Environmental Science and Policy*, October 2010.

Sylvain Charlebois, "There's a New World Order in Food," *Toronto Globe & Mail*, August 19, 2010.

Catherine M. Cooney, "The Perception Factor: Climate Change Gets Personal," *Environmental Health Perspectives*, November 1, 2010.

Daniel F. Doak and William F. Morris, "Demographic Compensation and Tipping Points in Climate-Induced Range Shifts," *Nature*, October 21, 2010.

Patrick Evans, "Milking the Planet Dry," *Geographical*, April 2009.

Hans-Martin Fussel, "How Inequitable Is the Global Distribution of Responsibility, Capability, and Vulnerability to Climate Change," *Global Environmental Change*, October 2010.

Justin Gillis, "In Weather Chaos, a Case for Global Warming," *New York Times*, August 14, 2010.

Guttmacher Institute, "Facts on Satisfying the Need for Contraception in Developing Countries," April 2010. www.guttmacher.org.

Kelly Hearn, "Climate May Heat Crises, Too, Military Analysts Say," *Christian Science Monitor*, March 18, 2009.

James R. Lee, "Global Warming Is Just the Tip of the Iceberg," *Washington Post*, January 4, 2009.

Claum MacLeod, "In Mongolia, an Unfolding Disaster," *USA Today*, February 25, 2010.

Robert McLeman, "Impacts of Population Change on Vulnerability and the Capacity to Adapt to Climate Change and Variability," *Population & Environment*, May 2010.

Anthony J. McMichael, "Paleoclimate and Bubonic Plague: A Forewarning of Future Risk?" *BMC Biology*, August 27, 2010.

Peter Miller, "It Starts at Home: We Already Know the Fastest, Least Expensive Way to Slow Climate Change: Use Less Energy," *National Geographic*, March 2009.

Sarah Murray, "Production Must Rise to Banish Hunger," *Financial Times*, October 15, 2010.

Vera Pavlova, "Moscow, Through a Cloud of Smoke," *New York Times*, August 15, 2010.

Population Reference Bureau, "2010 World Population Data Sheet." www.prb.org.

Clionadh Raleigh and Henrik Urdal, "Climate Change, Demography, Environmental Degradation, and Armed Conflict," Woodrow Wilson International Center for Scholars, 2007. www.wilsoncenter.org.

Doyle Rice, "Billions and Billions Served: How Population Affects Climate Change," *USA Today*, October 11, 2010.

Semil Shah, Sarath Guttikunda, and Ramesh P. Singh, "Himalayan Climate Change Threatens Regional Stability. Can India Help?" *Christian Science Monitor*, October 22, 2010.

Dan Smith and Janani Vivekananda, *Climate Change, Conflict and Fragility*, International Alert, November 2009. www.international-alert.org.

Kerri Smith, "The Population Problem," *Nature Reports Climate Change*, June 2008.

*Space Daily*, "Developing World's Crops Under Increased Threat," December 5, 2006.

Tanya Tillett, "Temperatures Rising: Sprawling Cities Have the Most Very Hot Days," *Environmental Health Perspectives*, October 2010.

Patrick Tucker, "Designing Buildings for Climate Change," *The Futurist*, September-October 2010.

UNESCO/World Water Assessment Program, *The Third United Nations World Water Development Report: Water in a Changing World*, 2009. www.unesco.org.

US Global Change Research Program, *Global Climate Change Impacts in the United States*, 2009. www.globalchange.gov.

Mike Williams, "Acting Like Neanderthals," *History Today*, October 2010.

Worldwatch Institute, *State of the World Population 2009*. www .worldwatch.org.

## Web sites

Population Action International (www.populationaction.org). Population Action International, which advocates for family planning and reproductive health around the world, is part of a collaboration with the National Center for Atmospheric Research and the Joint Global Change Research Institute to examine the intersection between climate change and population. The agency's Web site provides reports, data, maps, and other types of information about climate change and population.

Refugee Studies Center (www.rsc.ox.ac.uk). The Refugee Studies Center is a multidisciplinary center for research and teaching on the causes and consequences of forced migration. The center's Web site provides policy briefs and working papers, as well as free access to *Forced Migration Review*, a monthly publication that focuses on the issues of refugees and internally displaced people.

United Nations Environment Program (www.unep.org). UNEP's climate change Web site provides in-depth information on climate change adaptation and mitigation activities occurring around the world. Additionally, the Web site is a source of information about the UN Framework Convention on Climate Change (UNFCCC) and Conference of the Parties (COP) meetings.

United Nations Population Fund (www.unfpa.org). The United Nations Population Fund (UNFPA) is an international agency supporting countries as they seek to reduce poverty and increase the quality of life for men, women, and children around the world. UNFPA's annual "State of the World Population" reports and many others on population and climate change are available on the agency's Web site.

World Resources Institute (www.wri.org ). The World Resources Institute is an environmental think tank. The institute's Web site provides myriad reports, data, and statistics about climate change and population.

# Index

# About the Author

Jacqueline Langwith conducts policy research for the nonpartisan Michigan Legislative Service Bureau in the areas of biotechnology, health care, controlled substances, telecommunications, climate change, alternative and renewable energy, and the generation of electric power. She has a masters degree in biochemistry obtained from Michigan Technological University (MTU). Prior to her position at the bureau, Ms. Langwith managed the Plant Biotechnology Research Center at MTU, where she contributed to several research articles concerning the biosynthesis of lignin in woody plants, in publications such as the *Proceedings of the National Academy of Sciences*, *The Plant Cell*, and *Nature Biotechnology*.